The Story of the Livin

A Review of the Conclusions of Modern Biology in Regardto the Mechanism Which Controls the Phenomena of LivingActivity

H. W. Conn

Alpha Editions

This edition published in 2024

ISBN : 9789362997425

Design and Setting By
Alpha Editions
www.alphaedis.com
Email - info@alphaedis.com

As per information held with us this book is in Public Domain.
This book is a reproduction of an important historical work. Alpha Editions uses the best technology to reproduce historical work in the same manner it was first published to preserve its original nature. Any marks or number seen are left intentionally to preserve its true form.

Contents

PREFACE. .... - 1 -

INTRODUCTION. .... - 2 -

PART I. THE RUNNING OF THE
LIVING MACHINE. .... - 13 -

CHAPTER I. IS THE BODY A
MACHINE? .... - 15 -

CHAPTER II. THE CELL AND
PROTOPLASM. .... - 37 -

PART II. THE BUILDING OF THE
LIVING MACHINE. .... - 87 -

CHAPTER III. THE FACTORS
CONCERNED IN THE BUILDING OF
THE LIVING MACHINE. .... - 89 -

PREFACE.

That the living body is a machine is a statement that is frequently made without any very accurate idea as to what it means. On the one hand it is made with a belief that a strict comparison can be made between the body and an ordinary, artificial machine, and that living beings are thus reduced to simple mechanisms; on the other hand it is made loosely, without any special thought as to its significance, and certainly with no conception that it reduces life to a mechanism. The conclusion that the living body is a machine, involving as it does a mechanical conception of life, is one of most extreme philosophical importance, and no one interested in the philosophical conception of nature can fail to have an interest in this problem of the strict accuracy of the statement that the body is a machine. Doubtless the complete story of the living machine can not yet be told; but the studies of the last fifty years have brought us so far along the road toward its completion that a review of the progress made and a glance at the yet unexplored realms and unanswered questions will be profitable. For this purpose this work is designed, with the hope that it may give a clear idea of the trend of recent biological science and of the advances made toward the solution of the problem of life.

MIDDLETOWN, CONN., U.S.A.

October 1, 1898.

INTRODUCTION.

Biology a New Science.—In recent years biology has been spoken of as a new science. Thirty years ago departments of biology were practically unknown in educational institutions. To-day none of our higher institutions of learning considers itself equipped without such a department. This seems to be somewhat strange. Biology is simply the study of living things; and living nature has been studied as long as mankind has studied anything. Even Aristotle, four hundred years before Christ, classified living things. From this foundation down through the centuries living phenomena have received constant attention. Recent centuries have paid more attention to living things than to any other objects in nature. Linnæus erected his systems of classification before modern chemistry came into existence; the systematic study of zoology antedated that of physics; and long before geology had been conceived in its modern form, the animal and vegetable kingdoms had been comprehended in a scientific system. How, then, can biology be called a new science When it is older than all the others?

There must be some reason why this, the oldest of all, has been recently called a new science, and some explanation of the fact that it has only recently advanced to form a distinct department in our educational system. The reason is not difficult to find. Biology is a new science, not because the objects it studies are new, but because it has adopted a new relation to those objects and is studying them from a new standpoint. Animals and plants have been studied long enough, but not as we now study them. Perhaps the new attitude adopted toward living nature may be tersely expressed by saying that in the past it has been studied as at rest, while to-day it is studied as in motion. The older zoologists and botanists confined themselves largely to the study of animals and plants simply as so many museum specimens to be arranged on shelves with appropriate names. The modern biologist is studying these same objects as intensely active beings and as parts of an ever-changing history. To the student of natural history fifty years ago, animals and plants were objects to be classified; to the biologist of to-day, they are objects to be explained.

To understand this new attitude, a brief review of the history of the fundamental features of philosophical thought will be necessary. When, long ago, man began to think upon the phenomena of nature, he was able to understand almost nothing. In his inability to comprehend the activities going on around him he came to regard the forces of nature as manifestations of some supernatural beings. This was eminently natural. He had a direct consciousness of his own power to act, and it was natural for him to assume that the activities going on around him were caused by similar powers on the part of some being like himself, only superior to him. Thus he came to fill the unseen universe with gods controlling the forces of nature. The wind was the breath of one god, and the lightning a bolt thrown from the hands of another.

With advancing thought the ideas of polytheism later gave place to the nobler conception of monotheism. But for a long time yet the same ideas of the supernatural, as related to the natural, retained their place in man's philosophy. Those phenomena which he thought he could understand were looked upon as natural, while those which he could not understand were looked upon as supernatural, and as produced by the direct personal activity of some divine agency. As the centuries passed, and man's power of observation became keener and his thinking more logical, many of the hitherto mysterious phenomena became intelligible and subject to simple explanations. As fast as this occurred these phenomena were unconsciously taken from the realm of the supernatural and placed among natural phenomena which could be explained by natural laws. Among the first mysteries to be thus comprehended by natural law were those of astronomy. The complicated and yet harmonious motions of the heavenly bodies had hitherto been inexplicable. To explain them many a sublime conception of almighty power had arisen, and the study of the heavenly bodies ever gave rise to the highest thoughts of Deity. But Newton's law of gravitation reduced the whole to the greatest simplicity. Through the law and force of gravitation these mysteries were brought within the grasp of human understanding. They ceased to be looked upon as supernatural, and became natural phenomena as soon as the force of gravitation was accepted as a part of nature.

In other branches of natural phenomena the same history followed. The forces and laws of chemical affinity were formulated and studied, and physical laws and forces were comprehended. As these natural forces were grasped it became, little by little, evident that the various phenomena of nature were simply the result of nature's forces acting in accordance with nature's laws. Phenomena hitherto mysterious were one after another brought within the realm of law, and as this occurred a smaller and smaller portion of them were left within the realm of the so-called supernatural. By the middle of this century this advance had reached a point where scientists, at least, were ready to believe that nature's forces were all-powerful to account for nature's phenomena. Science had passed from the reign of mysticism to the reign of law.

But after chemistry and physics, with all the forces that they could muster, had exhausted their powers in explaining natural phenomena, there apparently remained one class of facts which was still left in the realm of the supernatural and the unexplained. The phenomena associated with living things remained nearly as mysterious as ever. Life appeared to be the most inexplicable phenomena of nature, and none of the forces and laws which had been found sufficient to account for other departments of nature appeared to have much influence in rendering intelligible the phenomena of life. Living organisms appeared to be actuated by an entirely unique force. Their shapes and structure showed so many marvellous adaptations to their surroundings as to render it apparently certain that their adjustment must have been the result of some intelligent planning, and not the outcome of blind force. Who could look upon the adaptation of the eye to light without seeing in It the result of intelligent design? Adaptation to conditions is seen in all animals and plants. These organisms are evidently complicated machines with their parts intricately adapted to each other and to surrounding conditions. Apart from animals and plants the only other similarly adjusted machines are those which have been made by human intelligence; and the inference seemed to be clear that a similar intelligence was needed to account for the living machine. The blind action of physical forces seemed inadequate. Thus the phenomena of life, which had been studied longer than any other phase of nature, continued to stand aloof from the rest and refused to fall into line with the general drift of thought. The living world seemed to give no promise of being included among natural phenomena, but still persisted in retaining its supernatural aspect.

It is the attempt to explain the phenomena of the living world by the same kind of natural forces that have been adequate to account for other phenomena, that has created modern Biology. So long as students simply studied animals and plants as objects for classification, as museum objects, or as objects which had been stationary in the history of nature, so long were they simply following along the same lines in which their predecessors had been travelling. But when once they began to ask if living nature were not perhaps subject to an intelligent explanation, to study living things as part of a general history and to look upon them as active moving objects whose motion and whose history might perhaps be accounted for, then at once was created a new department of thought and a new science inaugurated.

Historical Geology.—Preparation had been made for this new method of studying life by the formulation of a number of important scientific discoveries. Prominent among these stood historical geology. That the earth had left a record of her history in the rocks in language plain enough to be read appears to have been impressed upon scientists in the last of the century. That the earth has had a history and that man could read it became more and more thoroughly understood as the first decades of this century passed. The reading of that history proved a somewhat difficult task. It was written in a strange language, and it required many years to discover the key to the record. But under the influence of the writings of Lyell, just before the middle of the century, it began to appear that the key to this language is to be found by simply opening the eyes and observing what is going on around us to-day. A more extraordinary and more important discovery has hardly ever been made, for it contained the foundation of nearly all scientific discoveries which have been made since. This discovery proclaimed that an application of the forces still at work to-day on the earth's surface, but continued throughout long ages, will furnish the interpretation of the history written in the rocks, and thus an explanation of the history of the earth itself. The slow elevation of the earth's crust, such as is still going on to-day, would, if continued, produce mountains; and the washing away of the land by rains and floods, such as we see all around us, would, if continued through the long centuries, produce the valleys and gorges which so astound us. The explanation of the past is to be found in the present. But this geological history told of a history of life as well as a history of rocks. The history of the rocks has indeed been bound up in the history of life, and no sooner did it appear that the earth's crust has had a readable history than it appeared that living nature had a parallel history. If the present is a key to the past in interpreting geological history, should not the same be true of this history of life? It was inevitable that problems of life should come to the front, and that the study of life from the dynamical standpoint, rather than a statical, should ensue. Modern biology was the child of historical geology.

But historical geology alone could never have led to the dynamical phase of modern biology. Three other conceptions have contributed in an even greater degree to the development of this science.

Conservation of Energy.—The first of these was the doctrine of conservation of energy and the correlation of forces. This doctrine is really quite simple, and may be outlined as follows: In the universe, as we know it, there exists a certain amount of energy or power of doing work. This amount of energy can neither be increased nor decreased; energy can no more be created or destroyed than matter. It exists, however, in a variety of forms, which may be either active or passive. In the active state it takes some form of motion. The various forces which we recognize in nature—heat, light, electricity, chemism, etc.—are simply forms of motion, and thus forms of this energy. These various types of energy, being only expressions of the universal energy, are convertible into each other in such a way that when one disappears another appears. A cannon ball flying through the air exhibits energy of motion; but it strikes an obstacle and stops. The motion has apparently stopped, but an examination shows that this is not the case. The cannon ball and the object it strikes have been heated, and thus the motion of the ball has simply been transformed into a different form of motion, which we call heat. Or, again, the heat set free under the locomotive boiler is converted by machinery into the motion of the locomotive. By still different mechanism it may be converted into electric force. All forms of motion are readily convertible into each other, and each form in which energy appears is only a phase of the total energy of nature.

A second condition of energy is energy at rest, or potential energy. A stone on the roof of a house is at rest, but by virtue of its position it has a certain amount of potential energy, since, if dislodged, it will fall to the ground, and thus develop energy of motion. Moreover, it required to raise the stone to the roof the expenditure of an amount of energy exactly equal to that which will reappear if the stone is allowed to fall to the ground. So in a chemical molecule, like fat, there is a store of potential energy which may be made active by simply breaking the molecule to pieces and setting it free. This occurs when the fat burns and the energy is liberated as heat. But it required at some time the expenditure of an equal amount of energy to make the molecule. When the molecule of fat was built in the plant which produced it, there was used in its construction an amount of solar energy exactly equivalent to the energy which may be liberated by breaking the molecule to pieces. The total sum of the active and potential energy in the universe is thus at all times the same.

This magnificent conception has become the cornerstone of modern science. As soon as conceived it brought at once within its grasp all forms of energy in nature. It is primarily a physical doctrine, and has been developed chiefly in connection with the physical sciences. But it shows at once a possible connection between living and non-living nature. The living organism also exhibits motion and heat, and, if the doctrine of the conservation of energy be true, this energy must be correlated with other forms of energy. Here is a suggestion that the same laws control the living and the non-living world; and a suspicion that if we can find a natural explanation of the burning of a piece of coal and the motion of a locomotive, so, too, we may find a natural explanation of the motion of a living machine.

Evolution—A second conception, whose influence upon-the development of biology was even greater, was the doctrine of evolution. It is true that the doctrine of evolution was no new doctrine with the middle of this century, for it had been conceived somewhat vaguely before. But until historical geology had been formulated, and until the idea of the unity of nature had dawned upon the minds of scientists, the doctrine of evolution had little significance. It made little difference in our philosophy whether the living organisms were regarded as independent creations or as descended from each other, so long as they were looked upon as a distinct realm of nature without connection with the rest of nature's activity. If they are distinct from the rest of nature, and therefore require a distinct origin, it makes little difference whether we looked upon that origin as a single originating point or as thousands of independent creations. But so soon as it appeared that the present condition of the earth's crust was formed by the action of forces still in existence, and so soon as it appeared that the forces outside of living forces, including astronomical, physical and chemical forces, are all correlated with each other as parts of the same store of energy, then the problem of the origin of living things assumed a new meaning. Living things became then a part of nature, and demanded to be included in the same general category. The reign of law, which was claiming that all nature's phenomena are the result of natural rather than supernatural powers, demanded some explanation of the origin of living things. Consequently, when Darwin pointed out a possible way in which living phenomena could thus be included in the realm of natural law, science was ready and anxious to receive his explanation.

Cytology.—A third conception which contributed to the formulation of modern biology was derived from the facts discovered in connection with the organic cell and protoplasm. The significance of these facts we shall notice later, but here we may simply state that these discoveries offered to students simplicity in the place of complexity. The doctrine of cells and protoplasm appeared to offer to biologists no longer the complicated problems which were associated with animals and plants, but the same problems stripped of all side issues and reduced to their lowest terms. This simplifying of the problems proved to be an extraordinary stimulus to the students who were trying to find some way of understanding life.

New Aspects of Biology.—These three conceptions seized hold of the scientific world at periods not very distant from each other, and their influence upon the study of living nature was immediate and extraordinary. Living things now came to be looked upon not simply as objects to be catalogued, but as objects which had a history, and a history which was of interest not merely in itself, but as a part of a general plan. They were no longer studied as stationary, but as moving phases of nature. Animals were no longer looked upon simply as beings now existing, but as the results of the action of past forces and as the foundation of a different series of beings in the future. The present existing animals and plants came to be regarded simply as a step in the long history of the universe. It appeared at once that the study of the present forms of life would offer us a means of interpreting the past and perhaps predicting the future.

In a short time the entire attitude which the student assumed toward living phenomena had changed. Biological science assumed new guises and adopted new methods. Even the problems which it tried to solve were radically changed. Hitherto the attempt had been made to find instances of purpose in nature. The marvellous adaptations of living beings to their conditions had long been felt, and the study of the purposes of these adaptations had inspired many a magnificent conception. But now the scientist lost sight of the purpose in hunting for the cause. Natural law is blind and can have no purpose. To the scientist, filled with the thought of the reign of law, purpose could not exist in nature. Only cause and effect appeal to him. The present phenomena are the result of forces acting in the past, and the scientist's search should be not for the purpose of an adaptation, but for the action of the forces which produced it. To discover the forces and laws which led to the development of the present forms of animals and plants, to explain the method by which these forces of nature have acted to bring about present results, these became the objects of scientific research. It no longer had any meaning to find that a special organ was adapted to its conditions; but it was necessary to find out how it became adapted. The difference in the attitude of these two points of view is world-wide. The former fixes the attention upon the end, the latter upon the means by which the end was attained; the former is what we sometimes call teleological, the latter scientific; the former was the attitude of the study of animals and plants before the middle of this century, the latter the spirit which actuates modern biology.

The Mechanical Nature of Living Organisms.—This new attitude forced many new problems to the front. Foremost among them and fundamental to them all were the questions as to the mechanical nature of living organisms. The law of the correlation of force told that the various forms of energy which appear around us—light, heat, electricity, etc.—are all parts of one common store of energy and convertible into each other. The question whether vital energy is in like manner correlated with other forms of energy was now extremely significant. Living forces had been considered as standing apart from the rest of nature. Vital force, or vitality, had been thought of as something distinct in itself; and that there was any measurable relation between the powers of the living organism and the forces of heat and chemical affinity was of course unthinkable before the formulation of the doctrine of the correlation of forces. But as soon as that doctrine was understood it began to appear at once that, to a certain extent at least, the living body might be compared to a machine whose function is simply to convert one kind of energy into another. A steam engine is fed with fuel. In that fuel is a store of energy deposited there perhaps centuries ago. The rays of the sun, shining on the world in earlier ages, were seized upon by the growing plants and stored away in a potential form in the wood which later became coal. This coal is placed in the furnace of the steam engine and is broken to pieces so that it can no longer hold its store of energy, which is at once liberated in its active form as heat. The engine then takes the energy thus liberated, and as a result of its peculiar mechanism converts it into the motion of its great fly-wheel. With this notion clearly in mind the question forces itself to the front whether the same facts are not true of the living animal organism. It, too, is fed with food containing a store of energy; and should we not regard it, like the steam engine, simply a machine for converting this potential energy into motion, heat, or some other active form? This problem of the correlation of vital and physical forces is inevitably forced upon us with the doctrine of the correlation of forces. Plainly, however, such questions were inconceivable before about the middle of the nineteenth century.

This mechanical conception of living activity was carried even farther. Under the lead of Huxley there arose in the seventh decade of the century a view of life which reduced it to a pure mechanism. The microscope had, at that time, just disclosed the universal presence in living things of that wonderful substance, protoplasm. This material appeared to be a homogeneous substance, and a chemical study showed it to be made of chemical elements united in such a way as to show close relation to albumens. It appeared to be somewhat more complex than ordinary albumen, but it was looked upon as a definite chemical compound, or, perhaps, as a simple mixture of compounds. Chemists had shown that the properties of compounds vary with their composition, and that the more complex the compound the more varied its properties. It was a natural conception, therefore, that protoplasm was a complex chemical compound, and that its vital properties were simply the chemical properties resulting from its composition. Just as water possesses the power of becoming solid at certain temperatures, so protoplasm possesses the power of assimilating food and growing; and, since we do not doubt that the properties of water are the result of its chemical composition, so we may also assume that the vital properties of protoplasm are the result of its chemical composition. It followed from this conclusion that if chemists ever succeeded in manufacturing the chemical compound, protoplasm, it would be alive. Vital phenomena were thus reduced to chemical and mechanical problems.

These ideas arose shortly after the middle of the century, and have dominated the development of biological science up to the present time. It is evident that the aim of biological study must be to test these conceptions and carry them out into details. The chemical and mechanical laws of nature must be applied to vital phenomena in order to see whether they can furnish a satisfactory explanation of life. Are the laws and forces of chemistry sufficient to explain digestion? Are the laws of electricity applicable to an understanding of nervous phenomena? Are physical and chemical forces together sufficient to explain life? Can the animal body be properly regarded as a machine controlled by mechanical laws? Or, on the other hand, are there some phases of life which the forces of chemistry and physics cannot account for? Are there limits to the application of natural law to explain life? Can there be found something connected with living beings which is force but not correlated with the ordinary forms of energy? Is there such a thing as vital energy, or is the so-called vital force simply a name which we have given to the peculiar manifestations of ordinary energy as shown in the substance protoplasm? These are some of the questions that modern biology is trying to answer, and it is the existence of such questions which has made modern biology a new science. Such questions not only did not, but could not, have arisen before the doctrines of the conservation of energy and evolution had made their impression upon the thought of the world.

Significance of the New Biological Problems—It is further evident that the answers to these questions will have a significance reaching beyond the domain of biology proper and affecting the fundamental philosophy of nature. The answer will determine whether or not we can accept in entirety the doctrines of the conservation of energy and evolution. Plainly if it should be found that the energy of animate nature was not correlated with other forms of energy, this would demand either a rejection or a complete modification of our doctrine of the conservation of energy. If an animal can create any energy within itself, or can destroy any energy, we can no longer regard the amount of energy of the universe as constant. Even if that subtile form of force which we call nervous energy should prove to be uncorrelated with other forms of energy, the idea of the conservation of energy must be changed. It is even possible that we must insist that the still more subtile form of force, mental force, must be brought within the scope of this great law in order that it be implicitly accepted. This law has proved itself strictly applicable to the inanimate world, and has then thrust upon us the various questions in regard to vital force, and we must recognize that the real significance of this great law must rest upon the possibility of its application to vital phenomena.

No less intimate is the relation of these problems to the doctrine of evolution. Evolution tries to account for each moment in the history of the world as the result of the conditions of the moment before. Such a theory loses its meaning unless it can be shown that natural forces are sufficient to account for living phenomena. If the supernatural must be brought in here and there to account for living phenomena, then evolution ceases to have much meaning. It is undoubtedly a fact that the rapidly developing ideas along the above mentioned lines of dynamical biology have, been potent factors in bringing about the adoption of evolution. Certain it is that, had it been found that no correlation could be traced between vital and non-vital forces, the doctrine of evolution could not have stood, and even now the special significance which we shall in the end give to evolution will depend upon how we succeed in answering the questions above outlined. The fact is that this problem of the mechanical explanation of vital phenomena forms the capstone of the arch, the sides of which are built of the doctrines of the conservation of energy and the theory of evolution. To the presentation of these problems the following pages will be devoted. The fact that both the doctrine of the conservation of energy and that of evolution are practically everywhere accepted indicates that the mechanical nature of vital forces is regarded as proved. But there are still many questions which are not so easily answered. It will be our purpose in the following discussion to ascertain just what are these problems in dynamical biology and how far they have been answered. Our object will be then in brief to discover to what extent the conception of the living organism as a machine is borne out by the facts which have been collected in the last quarter century, and to learn where, if anywhere, limits have been found to our possibility of applying the forces of chemistry and physics to an explanation of life. In other words, we shall try to see how far we have been able to understand living phenomena in terms of natural force.

Outline of the Subject.—The subject, as thus presented, resolves itself at once into two parts. That the living organism is a machine is everywhere recognized, although some may still doubt as to the completeness of the comparison. In the attempt to explain the phenomena of life we have two entirely different problems. The first is manifestly to account for the existence of this machine, for such a completed piece of mechanism as a man or a tree cannot be explained as a result of simple accident, as the existence of a rough piece of rock might be explained. Its intricacy of parts and their purposeful interrelation demands explanation, and therefore the fundamental problem is to explain how this machine came into existence. The second problem is simpler, for it is simply to explain the running of the machine after it is made. If the organism is really a machine, we ought to be able to find some way of explaining its actions as we can those of a steam engine.

Of these two problems the first is the more fundamental, for if we fail to find an explanation for the existence of the machine, our explanation of its method of action is only partly satisfactory. But the second question is the simpler, and must be answered first. We cannot hope to explain the more puzzling matter of the origin of the machine unless we can first understand how it acts. In our treatment of the subject, therefore, we shall divide it into two parts:

> I. The Running of the Living Machine.
>
> II. The Origin of the Living Machine.

PART I.
THE RUNNING OF THE LIVING MACHINE.

CHAPTER I.
IS THE BODY A MACHINE?

The problem before us in this section is to find out to what extent animals and plants are machines. We wish to determine whether the laws and forces which regulate their activities are the same as the laws and forces with which we experiment in the chemical and physical laboratory, and whether the principles of mechanics and the doctrine of the conservation of energy apply equally well in the living machine and the steam engine.

It might be inferred that the proper method of study would be to confine our attention largely to the simplest forms of life, since the problems would be here less complicated, and therefore of easier solution. This, however, has not been nor can it be the method of study. Our knowledge of the processes of life have been derived largely from the most rather than the least complex forms. We have a better knowledge of the physiology of man and his allies than any other animals. The reason for this is plain enough. In the first place, there is a value in the knowledge of the life activities of man entirely apart from any theoretical aspects, and hence human physiology has demanded attention for its own sake. The practical utility of human physiology has stimulated its study for centuries; and in the last fifty years of scientific progress it has been human physiology and that of allied animals that has attracted the chief attention of physiologists. The result is that while the physiology of man is tolerably well known, that of other animals is less understood the farther we get away from man and his allies. For this reason most of our knowledge of the living body as a machine must be derived from the study of man. This is, however, fortunate rather than otherwise. In the first place, it enables us to proceed from the known to the unknown; and in the second place, more interest attaches to the problem as connected with human physiology than along any other line. In our discussion, therefore, we shall refer chiefly to the physiology of man. If we find that the functions of human life are amenable to a mechanical explanation we cannot hesitate to believe that this will be equally true of the lower orders of nature. For similar reasons little reference will be made to the mechanism of plant life. The structure of the plant is simpler and its activities are much more easily referable to mechanical principles than are those of animals. For these reasons it will only be necessary for us to turn our attention to the life activities of the higher animals.

What is a Machine?—Turning now to our more immediate subject of the accuracy of the statement that the body is a machine, we must first ask what is meant by a machine? A brief definition of a machine might be as follows: A machine is a piece of apparatus so designed that it can change one kind of energy into another for a definite purpose. Energy, as already noticed, is the power of doing work, and its ordinary active forms are heat, motion, electricity, light, etc.; but it may be in a passive or potential form, and in this form stored within a chemical molecule. These various forms of energy are readily convertible into each other; and any form of apparatus designed for the purpose of producing such a conversion is called a machine. A dynamo is thus a machine so adjusted that when mechanical motion is supplied to it the energy of motion is converted into electricity; while an electromotor, on the other hand, is a piece of apparatus so designed that when electricity is applied to it, it is converted into motion. A steam engine, again, is designed to convert potential or passive energy into active energy. Potential energy in the form of chemical composition (coal) is supplied to the engine, and this energy is first liberated in the active form of heat and then is converted into the motion of the great fly-wheel. In all these cases there is no energy or power created, for the machine must be always supplied with an amount of energy equal to that which it gives back in another form. Indeed, a larger amount of energy must be furnished the machine than is expected back, for there is always an actual loss of available energy. In the process of the conversion of one form of energy into another some of the energy, from friction or other cause, takes the form of heat, and is then radiated into space beyond our reach. It is, of course, not destroyed, for energy cannot be destroyed; but it has assumed a form called radiant heat, which is not available for our uses. A machine thus neither creates nor destroys energy. It receives it in one form and gives it back in another form, with an inevitable loss of a portion of the energy as radiant heat. With this understanding, we may now ask if the living body can be properly compared with a machine.

A General Comparison of a Body and a Machine.--That the living body exhibits the ordinary types of energy is of course clear enough when we remember that it is always in motion and is always radiating heat—two of the most common types of physical energy. That this energy is supplied to the body as it is to other machines, in the form of the energy of chemical composition, will also need no further proof when it is remembered that it is necessary to supply the body with appropriate food in order that it may do work. The food we eat, like coal, represents so much solar energy which is stored up by the agency of plant life, and the close comparison between feeding the body to enable it to work and feeding the engine to enable it to develop energy is so evident that it demands no further demonstration. The details of the problem may, however, present some difficulties.

The first question which presents itself is whether the only power the body possesses is, as in the case with other machines, to transform energy without being able to create or destroy it? Can every bit of energy shown by the living organism be accounted for by energy furnished in the food, and conversely can all the energy furnished in the food be found manifested in the living organism?

The theoretical answer to this question in terms of the law of the conservation of energy is clear enough, but it is by no means so easy to answer it by experimental data. To obtain experimental demonstration it would be necessary to make an accurate determination of the amount of energy an individual receives during a given period, and at the same time a similar measurement of the amount of energy liberated in his body either as motion or heat. If the body is a machine, these two should exactly balance, and if they do not balance it would indicate that the living organism either creates or destroys energy, and is therefore not a machine. Such experiments are exceedingly difficult. They must be performed usually upon man rather than other animals, and it is necessary to inclose an individual in an absolutely sealed space with arrangements for furnishing him with air and food in measured quantity, and with appliances for measuring accurately the work he does and the heat given off from his body. In addition, it is necessary to measure the exact amount of material he eliminates in the form of carbonic acid and other excretions. Such experiments present many difficulties which have not yet been thoroughly overcome, but they have been attempted by several investigators. For the purpose of such an experiment scientists have allowed themselves to be shut up in a small chamber six or eight feet in length, in which their only communication with the outer world is by telephone and through a small opening in the side of the chamber, occasionally opened for a second or two to supply the prisoner with food. In such a chamber they have remained as long as twelve days. In these experiments it is necessary to take account not only of the food eaten, but of the actual amount of this food which is used by the body. If the person gains in weight, this must mean that he is storing up in his body material for future use; while if he loses in weight, this means that he is consuming his own tissues for fuel. Careful daily records of his weight must therefore be taken. Estimates of the solids, liquids, and gases given off from his body must be obtained, for to carry out the experiment an exact balance must be made between the income and the outgo. The apparatus devised for such experiments has been made very delicate; so delicate, indeed, that the rising of the individual in the box from his chair is immediately seen in a rise in temperature of the apparatus. But even with this delicacy the apparatus is comparatively coarse, and can measure only the most apparent forms of energy. The more subtle types of energy, such as nervous force, if this is to be regarded as energy, do not make any impression on the apparatus.

The obstacles in the way of these experiments do not particularly concern us, but the general results are of the greatest significance for our purpose. While, for manifest reasons, it has not been possible to carry on these experiments for any great length of time, and while the results have not yet been very accurately refined, they are all of one kind and teach unhesitatingly one conclusion. So far as concerns measurable energy or measurable material, the body behaves just like any other machine. If the body is to do work in this respiration apparatus, it does so only by breaking to pieces a certain amount of food and using the energy thus liberated, and the amount of food needed is proportional to the amount of work done. When the individual simply walks across the floor, or even rises from his chair, this is accompanied by an increase in the amount of food material broken up and a consequent increase in the amount of refuse matter eliminated and the heat given off. The income and outgo of the body in both matter and energy is balanced. If, during the experimental period, it is found that less energy is liberated than that contained in the food assimilated, it is also found that the body has gained in weight, which simply means that the extra energy has been stored in the body for future use. No more energy can be obtained from the body than is furnished, and for all furnished in the food an equivalent amount is regained. There is no trace of any creation or destruction of energy. While, on account of the complexity of the experimenting, an absolutely strict balance sheet cannot be made, all the results are of the same nature. So far as concerns measurable energy, all the facts collected bear out the theoretical conception that the living body is to be regarded as a machine which converts the potential energy of chemical composition, stored passively in its food, into active energy of motion and heat.

It is found, however, that the body is a machine of a somewhat superior grade, since it is able to convert this potential energy into motion with less loss than the ordinary machine. As noticed above, in all machines a portion of the energy is converted into heat and rendered unavailable by radiating into space. In an ordinary engine only about one-fifteenth of the energy furnished in the coal can be regained in the form of motive power, the rest being radiated from the machine as heat. Some of our better engines to-day utilize a somewhat larger part, but most of them utilize less than one-tenth. The experiments with the living body in the respiration apparatus above described, give a means of determining the proportion of the energy furnished in the form of food which can be utilized in the form of motive force. This figure appears to be decidedly larger than that obtained by any machine yet devised by man.

The conclusion of the matter up to this point is then clear. If we leave out of account the phenomena of the nervous system, which we shall consider presently, the general income and outgo of the body as concerns matter and energy is such that the body must be regarded as a machine, which, like other machines, simply transforms energy without creating or destroying it. To this extent, at least, animals conform to the law of the conservation of energy and are veritable machines.

Details of the Action of the Machine.—We turn next to some of the subordinate problems concerning the details of the action of the living machine. We have a clear understanding of the method of action of a steam engine. Its mechanism is simple, and, moreover, it was designed by human intelligence. We can understand how the force of chemical affinity breaks up the chemical composition of the coal, how the heat thus liberated is applied to the water to vapourize it; how the vapour is collected in the boiler under pressure; how this pressure is applied to the piston in the cylinder, and how this finally results in the revolution of the fly-wheel. It is true that we do not understand the underlying forces of chemism, etc., but these forces certainly exist and are the foundation of science. But the mechanism of the engine is intelligible. Our understanding of it is such that, with the forces of chemistry and physics as a foundation, we can readily explain the running of the machine. Our next problem, therefore, is to see if we can in the same way reach an understanding of the phenomena of the living machine. Can we, by the use of these same chemical and physical forces, explain the activities taking place in the living organism? Can the motion of the body, for example, be made as intelligible as the motion of the steam engine?

Physical Explanation of the Chief Vital Functions.—The living machine is, of course, vastly more complicated than the steam engine, and there are many different processes which must be considered separately. There is not space in a work of this size to consider them all carefully, but we may select a few of the vital functions as illustrations of the method which is pursued. It will be assumed that the fundamental processes of human physiology are understood by the reader, and we shall try to interpret some of them in terms of chemical and physical force.

Digestion.—The first step in this transformation of fuel is the process of digestion. Now this process of digestion is nothing mysterious, nor does it involve any peculiar or special forces. Digestion of food is simply a chemical change therein. The food which is taken into the body in the form of sugar, starch, fat or protein, is acted upon by the digestive juices in such a way that its chemical nature is slightly changed. But the changes that thus occur are not peculiar to the living body, since they will take place equally well in the chemist's laboratory. They are simply changes in the molecular structure of the food material, and only such changes as are simple and familiar to the chemist. The forces which effect the change are undoubtedly those of chemical affinity. The only feature of the process which is not perfectly intelligible in terms of chemical law is the nature of the digestive juices. The digestive fluids of the mouth and stomach contain certain substances which possess a somewhat remarkable power, inasmuch as they are able to bring about the chemical changes which occur in the digestion of food. An example will make this clearer. One of the digestive processes is the conversion of starch into sugar. The relation of these two bodies is a very simple one, starch being readily converted into sugar by the addition to its molecule of a molecule of water. The change can not be produced by simply adding starch to water, but the water must be introduced into the starch molecule. This change can be brought about in a variety of ways, and is undoubtedly effected by the forces of chemical affinity. Chemists have found simple methods of producing this chemical union, and the manufacture of sugar out of starchy material has even become something of a commercial industry. One of the methods by which this change can be produced is by adding to the starch, along with some water, a little saliva. The saliva has the power of causing the chemical change to occur at once, and the molecule of water enters into the starch molecule and forms sugar. Now we do not understand how this saliva possesses this power to induce the chemical change. But apparently the process is of the simplest character and involves no greater mystery than chemical affinity. We know that the saliva contains a certain material called a ferment, which is the active agent in bringing about the change. This ferment is not alive, nor does it need any living environment for its action. It can be separated from the saliva in the form of a dry amorphous powder, and in this form can be preserved almost indefinitely, retaining its power to effect the change whenever put under proper conditions. The change of starch into sugar is thus a simple chemical change occurring under the influence of chemical affinity under certain conditions. One of the conditions is the presence of this saliva ferment. If we can not exactly understand how the ferment produces this action, neither do we exactly understand how a spark causes a bit of gunpowder to explode. But we can not doubt that the latter is a purely natural result of the relation of chemical and physical forces, and there is no more reason for doubting it in the former case.

What is true of the digestion of starch by saliva is equally true of the digestion of other foods in the stomach and intestine. Each of the digestive juices contains a ferment which brings about a chemical change in the food. The changes are always chemical changes and are the result of chemical forces. Apart from the presence of these ferments there is really little difference between laboratory chemistry and living chemistry.

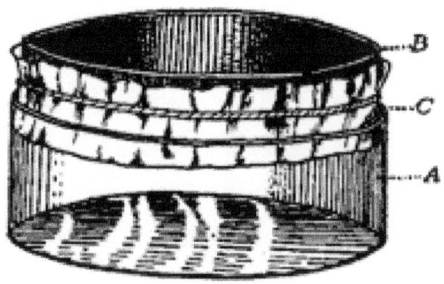

FIG. 1.—To illustrate osmosis. In the vessel A is a solution of sugar; in B is pure water. The two are separated by the mebrane C. The sugar passes through the membrane into B.

Absorption of food.—The next function of this machine to attract our attention is the absorption of food from the intestine into the blood. The digested food is carried down the alimentary canal in a purely mechanical fashion by muscular action, and when it reaches the intestine it begins to pass through its walls into the blood. In this absorption we find engaged another set of forces, the chief of which appears to be the physical force of osmosis. The force of osmosis has no special connection with life. If a membrane separates two liquids of different composition (Fig. i), a force is exerted on the liquids which cause them to pass through the membrane, each passing through the membrane into the other compartment. The force which drives these liquids through the membrane is considerable, and may sometimes be exerted against considerable pressure. A simple experiment will illustrate this force. In Fig. 2 is represented a membranous bag tightly fastened to a glass tube. The bag is filled with a strong solution of sugar, and is immersed in a vessel containing pure water. Under these conditions some of the sugar solution passes through the bag into the water, and some of the water passes from the vessel into the bag. But if the solution of sugar is inside the bag and the pure water outside, the amount of liquid passing into the bag is greater than the amount passing out; the bag soon becomes distended and the water even rises in the tube to a considerable height at a(Fig. 2). The force here concerned is a force known as osmosis or dialysis, and is always exerted when two different solutions of certain substances are separated from each other by a membrane. The substances in solution will, under these conditions, pass from the dense to the weaker solution. The process is a purely physical one.

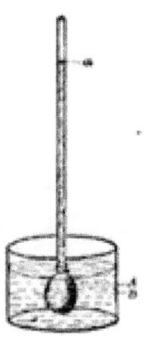

FIG. 2.—In the bladder
A is a sugar solution
In the vessel B
is pure water.
Sugar passes out
and water into
the bladder until it
rises in the tube
to a.

This process of osmosis lies at the basis of the absorption of food from the alimentary canal. In the first place, most of the food when swallowed is not soluble, and therefore not capable of osmosis. But the process of digestion, as we have seen, changes the chemical nature of the food. The food, as the result of chemical change, has become soluble, and after being dissolved it is dialyzable—i.e., capable of osmosis. After digestion, therefore, the food is dissolved in the liquids in the stomach and intestine, and is in proper condition for dialysis. Furthermore, the structure of the intestine is such as to produce conditions adapted for dialysis. This can be understood from Fig. 3, which represents diagrammatically a cross section through the intestinal wall. Within the intestinal wall, at A, is the food mass in solution. At B are shown little projections of the intestinal wall, called villi extending into this food and covered by a membrane. One of these villi is shown more highly magnified in Fig. 4, in which B shows this membrane. Inside of these villi are blood-vessels, C, and it will be thus seen that the membrane, B, separates two liquids, one containing the dissolved food outside the villus, and the other containing blood inside the villus. Here are proper conditions for osmosis, and this process of dialysis will take place whenever the intestinal contents holds more dialyzable material than the blood. Under these conditions, which will always occur after food has been digested by the digestive juices, the food will begin to pass through this membranous wall of the intestine into the blood under the influence of the physical force of osmosis. Thus the primary factor in food absorption is a physical one.

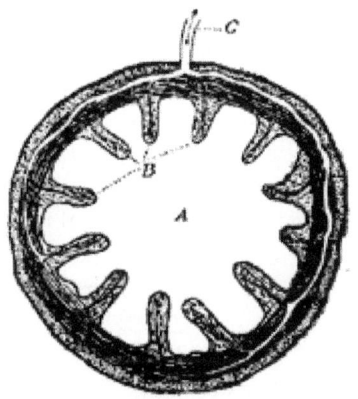

FIG. 3—Diagram of the intestinal walls. A, lumen of intestine filled with digested food. B, villi, containing blood vessels. C, larger blood vessel, which carries blood with absorbed food away from the intestine.

We must notice, however, that the physical force of osmosis is not the only factor concerned in absorption. In the first place, it is found that the food during its passage through the intestinal wall, or shortly afterwards, undergoes a further change, so that by the time it has fairly reached the blood it has again changed its chemical nature. These changes are, however, of a chemical nature, and, while we do not yet know very much about them, they are of the same sort as those of digestion, and involve probably nothing more than chemical processes.

Secondly, we notice that there is one phase of absorption which is still obscure. Part of the food is composed of fat, and this fat, as the result of digestion, is mechanically broken up into extremely minute droplets. Although these droplets are of microscopic size they are not actually in solution, and therefore not subject to the force of osmosis which only affects solutions. The osmotic force will not force fat drops through membranes, and to explain their passage through the walls of the intestine requires something additional. We are as yet, however, able to give only a partial explanation of this matter. The inner wall of the intestine is not an inert, lifeless membrane, but is made of active bits of living matter. These bits of living matter appear to seize hold of the droplets of oil by means of little processes which they thrust out, and then pass them through their own bodies to excrete them on their inner surface into the blood vessels. Fig. 5 shows a few of these living bits of the membrane, each containing several such fat droplets. This fat absorption thus appears to be a vital process, and not one simply controlled by physical forces like osmosis. Here our explanation runs against what we call vital power of the ultimate elements of the body. The consideration of this vital feature we must, of course, investigate further; but this will be done later. At present our purpose is a general comparison of the body and a machine, and we may for a little postpone the consideration of this vital phenomenon.

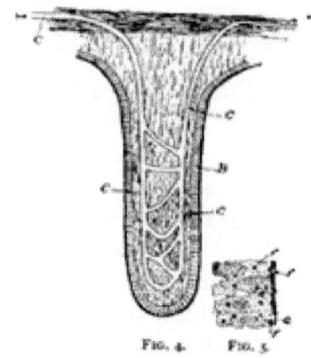

FIG. 4.—Diagram of a single villus enlarged.
B represents the membranous
surface covering the villus; C, the blood-vessels
within the villus.
FIG. 5.—An enlarged figure of four cells of
the membrane B in Fig. 4. The free
surface is at a; f shows fat droplets in
process of passage through the cells.

Circulation.—The next piece of mechanism for us to consider in this machine is the device for distributing this fuel to the various parts of the machine where it is to be used as a source of energy, corresponding in a sense to the fireman of a locomotive. This mechanism we call the circulatory system. It consists of a series of tubes, or blood vessels, running to every part of the body and supplying every bit of tissue. Within the tubes is the blood, which, from its liquid nature, is easily forced around the body through the tubes. At the centre of the system is a pump which keeps the blood in motion. The tubes form a closed system, such that the pump, or heart, may suck the blood in from one side to force it out into the tubes on the other side; and the blood, after passing over the body in this closed set of tubes, is finally brought back again to be forced once more over the same path. As this blood is carried around the body it conveys from one part of the machine to another all material that needs distribution. While in the intestine, as already noticed (Fig. 3), it receives the food, and now this food is carried by the circulation to the muscles or the other organs that need it. While in the lungs the blood receives oxygen, and this oxygen is then carried to those parts of the body that need it. The circulatory system is thus simply a medium by which each part of the machine may receive its proper share of the supplies needed for its action.

Now in this circulation we have again to do with chemical and physical forces. All of its general phenomena are based upon purely mechanical principles. The action of the heart—leaving out of consideration for a moment its muscular power—is that of a simple pump. It is provided with valves whose action is as simple and as easy to understand as those of any water pump. By the action of these valves the blood is kept circulating in one direction. The blood vessels are elastic, and the study of the effect of a liquid pumped rhythmically into elastic tubes explains with simplicity the various phenomena associated with the circulation. For example, the rhythmically contracting heart forces a small quantity of blood into the arteries at short intervals. These tubes are large near the heart, but smaller at their ends, where they flow into the veins, so that the blood does not flow out into the veins so readily as it flows in from the heart. The jet of blood that is sent in with every beat of the heart slightly stretches the artery, and the tension thus produced causes the blood to continue to flow between the beats. But the heart continues beating, and there is an accumulation of the blood in the arteries until it exists under some pressure—a pressure sufficient to force it rapidly through the small ends of the arteries into the veins. After passing into the veins the pressure is at once removed, since the veins are larger than the arteries, and there is no resistance to the flow of the blood. Hence the blood in the arteries is under pressure, while there is little or no pressure in the veins. Into the details of this matter we need not go, but this will be sufficient to indicate that the whole process is a mechanical one.

We must not fail to see, however, that in this problem of circulation there are two points at least where once more we meet with that class of phenomena which we still call vital. The beating of the heart is the first of these, for this is active muscular power. The second is a contraction of the smaller blood-vessels which regulates the blood supply. Both of these phenomena are phases of muscular activity, and will be included under the discussion of other similar phenomena later.

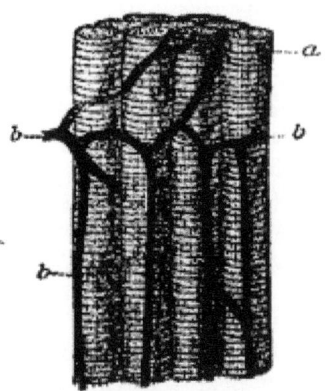

FIG. 6.—A bit of muscle with its blood-vessels:
a, the muscle fibres; b, the minute blood-vessels.
The fibres and vessels are bathed in lymph
(not shown in the figure), and food material passes through
the walls of the blood-vessels into this lymph.

We next notice that not only is the distribution of the blood explained upon mechanical principles, but the supplying of the active parts of the body with food is in the same way intelligible. As we have seen, the blood coming from the intestine contains the food material received from the digested food. Now when this blood in its circulation flows through the active tissues—for instance, the muscles—it is again placed under conditions where osmosis is sure to occur. In the muscles the thin-walled blood-vessels are surrounded and bathed by a liquid called lymph. Figure 6 shows a bit of muscle tissue, with its blood-vessels, which are surrounded by lymph. The lymph, which is not shown, fills all the space outside the blood-vessels, thus bathing both muscles and blood-vessels. Here again we have a membrane (i.e., the wall of the blood-vessel) separating two liquids, and since the lymph is of a different composition from the blood, dialysis between them is sure to occur, and the materials which passed into the blood in the intestine through the influence of the osmotic force, now pass out into the lymph under the influence of the same force. The food is thus brought into the lymph; and since the lymph lies in actual contact with the living muscle fibres, these fibres are now able to take directly from the lymph the material needed for their use. The power which enables the muscle fibre to take the material it needs, discarding the rest, is, again, one of the vital processes which we defer for a moment.

Respiration.—Pursuing the same line of study, we turn for a moment to the relation of the circulatory system to the function of supplying the body with oxygen gas. Oxygen is absolutely needed to carry on the functions of life; for these, like those of the engine, are based upon the oxidation of the fuel. The oxygen is derived from the air in the simplest manner. During its circulation the blood is brought for a fraction of a second into practical contact with air. This occurs in the lungs, where there are great numbers of air cells, in the walls of which the blood-vessels are distributed in great profusion. While the blood is in these vessels it is not indeed in actual contact with the air, but is separated from it by only a very thin membrane—so thin that it forms no hindrance to the interchange of gases. These air-cells are kept filled with air by simple muscular action. By the contraction of the muscles of the thorax the thoracic cavity is enlarged, and as a result air is sucked in in exactly the same way that it is sucked into a pair of bellows when expanded. Then the contraction of another set of muscles decreases the size of the thoracic cavity, and the air is squeezed out again. The action is just as truly mechanical as is that of the blacksmith's bellows.

The relation of the air to the blood is just as simple. In the blood there are various chemical ingredients, among which is one known as hæmoglobin. It does not concern us at present to ask where this material comes from, since this question is part of the broader question, the origin of the machine, to be discussed in the second part of this work. The hæmoglobin is a normal constituent of the blood, and, being red in colour, gives the red colour to the blood. This hæmoglobin has peculiar relations to oxygen. It can be separated from the blood and experimented upon by the chemist in his laboratory. It is found that when hæmoglobin is brought in contact with oxygen, under sufficient pressure it will form a chemical union with it. This chemical union is, however, what the chemist calls a loose combination, since it is readily broken up. If the oxygen is above a certain rather low pressure, the union will take place; while if the pressure be below this point the union is at once destroyed, and the oxygen leaves the hæmoglobin to become free. All of this is a purely chemical matter, and can be demonstrated at will in a test tube in the laboratory. But this union and disassociation is just what occurs as the foundation of respiration. The blood coming to the lungs contains hæmoglobin, and since the oxygen pressure in the air is quite high, this hæmoglobin unites at once with a quantity of oxygen while the blood is flowing through the air-vessels. The blood is then carried off in the circulation to the active tissues like the muscles. These tissues are constantly using oxygen to carry on their life processes, and consequently at all times use up about all the oxygen within their reach. The result is that in these tissues the oxygen pressure is very low, and when the oxygen-laden hæmoglobin reaches them the association of the hæmoglobin with oxygen is at once broken up and the oxygen set free in the tissue. It passes at once to the lymph, from which the active tissues seize it for the purpose of carrying on the oxidizing processes of the body. This whole matter of supplying the body with oxygen is thus fundamentally a chemical one, controlled by chemical laws.

Removal of Waste.—The next step in this life process is one of difficulty. After the food and oxygen have reached the tissues it is seized by the living cell. The food material is now oxidized by the oxygen and its latent energy is liberated, and appears in the form of motion or heat or some other vital function. Herein is the really mysterious part of the life process; but for the present we will overlook the mystery of this action, and consider the results from a purely material standpoint.

In a steam engine the fundamental process by which the latent energy of the fuel is liberated is that of oxidation. The oxygen of the air unites with the chemical elements of the fuel, and breaks up that fuel into simple compounds—which may be chiefly considered as three—carbonic dioxide (CO_2), water (H_2O), and ash. The energy contained in the original compound can not be held by these simpler bodies, and it therefore escapes as heat. Just the same process, with of course difference in details, is found in the living machine. The food, after reaching the living cell, is united with the oxygen, and, so far as chemical results are concerned, the process is much the same as if it occurred outside the body. The food is broken into simpler compounds and the contained energy is liberated. The energy is, by the mechanism of the machine, changed into motion or nervous impulse, etc. The food is broken into simple compounds, which are chiefly carbonic dioxide, water, and ash; the ash being, however, quite different from the ash obtained from burning coal. Now the engine must have its chimney to remove the gases and vapours (the CO_2 and H_2O) and its ashpit for the ashes. In the same way the living machine has its excretory system for removing wastes. In the removal of the carbonic acid and water we have to do once more with the respiratory system, and the process is simply a repetition of the story of gas diffusion, chemical union, and osmosis. It is sufficient here to say that the process is just as simple and as easily explained as those already described. The elimination of these wastes is simply a problem of chemistry and mechanics.

In the removal of the ash, however, we have something more, for here again we are brought up against the vital action of the cell. This ash takes chiefly the form of a compound known as urea, which finds its way into the general circulatory system. From the blood it is finally removed by the kidneys. In the kidneys are a large number of bits of living matter (kidney cells), which have the power of seizing hold of the urea as the blood is flowing over them, and after thus taking it out of the blood they deposit it in a series of tubes which lead to the bladder and hence to the exterior. The bringing of this ash to the kidney cell is a mechanical matter, based simply upon the flow of the blood. The seizing of the urea by the kidney cell is a vital phenomenon which we must waive for the moment.

Up to this point in the analysis there has been no difficulty, and no one can fail to agree with the conclusions. The position we reach is as follows: So far as relates to the general problems of energy in the universe the body is a machine. It neither creates nor destroys energy, but simply transforms one form into another. In attempting to explain the action of the machine, we find that for the functions thus far considered (sometimes called the vegetative functions) the laws of chemistry and physics furnish adequate explanation.

We must now look a little further, and question some of the functions the mechanical nature of which is less obvious. The whole operation thus far described is under the control of the nervous system, which acts somewhat like the engineer of an engine. Can this phase of living activity be included within the conception of the body as a machine?

Nervous System.—When we come to try to apply mechanical principles to the nervous system, we meet with what seems at first to be no thoroughfare. While dealing with the grosser questions of chemical compounds, heat, and motion, there is little difficulty in applying natural laws to the explanation of living phenomena. But the problem with the nervous system is very different. It is only to-day that we are finding that the problem is open to study, to say nothing of solution. It is true that mental and other nervous phenomena have been studied for a long time, but this study has been simply the study of these phenomena by themselves without a thought of their correlation with other phenomena of nature. It is a matter of quite recent conception that nervous phenomena have any direct relation to the other realms of nature.

Our first question must be whether we can find any correlation between nervous energy and other types of energy. For our purpose it will be convenient to distinguish between the phenomena of simple nervous transmission and the phenomena of mental activity. The former are the simpler, and offer the greatest hope of solution. If we are to find any correlation between nervous energy and other physical energy, we must do so by finding some way of measuring nervous energy and comparing it with the latter. This has been very difficult, for we have no way of measuring a nervous impulse directly. In the larger experiments upon the income and outgo of the body, in the respiration apparatus mentioned above, nervous phenomena apparently leave no trace. So far as experiments have gone as yet, there is no evidence of an expenditure of extra physical energy when the nervous system is in action. This is not surprising, however, for this apparatus is entirely too coarse to measure such delicate factors.

That there is a correlation between nervous energy and physical energy is, however, pretty definitely proved by experiments along different lines. The first step in this direction was to find that a nervous stimulus can be measured at least indirectly. When the nerve is stimulated there passes from one end to the other an impulse, and the rapidity with which it travels can be accurately measured. When such an impulse reaches the brain it may give rise to a conscious sensation, and a somewhat definite estimation can be made of the amount of time required for this. The periods are very short, of course, but they are not instantaneous. The nervous impulse, can be studied in still other ways. We find that the impulse can be started by ordinary forms of energy. A mechanical shock, a chemical or an electrical shock will develop nervous energy. Now these are ordinary forms of physical energy, and if, when they are applied to a nerve, they give rise to a nervous stimulus, the inference is certainly a legitimate one that the nerve is simply a bit of machinery adapted to the conversion of certain kinds of physical energy into nervous energy. If this is the case, then it is necessary to regard nervous energy as correlated with other forms of energy.

Other facts point in the same direction. Not only can the nervous stimulus be developed by an electric shock, but the strength of the stimulus is within certain limits proportional to the strength of the shock which produces it. Again, not only is it found that an electrical shock can develop a nervous stimulus, but conversely a nervous stimulus develops electrical energy. In ordinary nerves, even when not active, slight electric currents can be detected. They are extremely slight, and require the most delicate instruments for their detection. Now when a nerve is stimulated these currents are immediately affected in such a way that under proper conditions they are increased in intensity. The increase is sufficient to make itself easily seen by the motion of a galvanometer. The motion of the galvanometer under these conditions gives a ready means of studying the character of the nervous impulse. By its use it can be determined that the nerve impulse travels along the nerve like a wave, and we can approximately determine the length and shape of the wave and its relative height at various points.

Now what is the significance of all these facts for our discussion? Together they point clearly to the conclusion that nervous energy is correlated with other forms of physical energy. Since the nervous stimulus is started by other forms of energy, and since it can, in turn, modify ordinary forms of energy, we can not avoid the conclusion that the nervous impulse is only a special form of energy developed within the nerve. It is a form of wave motion peculiar to the nerve substance, but correlated with and developed from other types of energy. This, of course, makes the nerve simply a bit of machinery.

If this conclusion is true, the development of a nerve impulse would mean that a certain portion of food is broken to pieces in the body to liberate energy, and this should be accompanied by an elimination of carbonic dioxide and heat. This is easily shown to be true of muscle action. When we remove a muscle from the body it may remain capable of contracting for some time. By studying it under these conditions we find that it gives rise to carbonic dioxide and other substances, and liberates heat whenever it contracts. As already noticed, in the respiration experiments, whenever the individual experimented upon makes any motions, there is an accompanying elimination of waste products and a development of heat. But this does not appear to be demonstrable for the actions of the nervous system. Although very careful experiments have been made, it has as yet been found impossible to detect any rise in temperature when a nerve impulse is passing through a nerve, nor is there any demonstrable excretion of waste products. This would be a serious objection to the conception of the nerve as a machine were it not for the fact that the nerve is so small that the total sum of its nervous energy must be very slight. The total energy of this minute machine is so slight that it can not be detected by our comparatively rough instruments of measurement.

In short, all evidence goes to show that the nerve impulse is a form of motion, and hence of energy, correlated with other forms of physical energy. The nerve is, however, a very delicate machine, and its total amount of energy is very small. A tiny watch is a more delicate machine than a water-wheel, and its actions are more dependent upon the accuracy of its adjustment. The water-wheel may be made very coarse and yet be perfectly efficacious, while the watch must be fashioned with extreme delicacy. Yet the water-wheel transforms vastly more energy than the watch. It may drive the many machines in a factory, while the watch can do no more than move itself. But who can doubt that the watch, as well as the water-wheel, is governed by the law of the correlation of forces? So the nervous system of the living machine is delicately adjusted and easily put out of order, and its action involves only a small amount of energy; but it is just as truly subject to the law of the conservation of energy as is the more massive muscle.

Sensations.—Pursuing this subject further, we next notice that it is possible to trace a connection between physical energy and sensations. Sensations are excited by certain external forms of motion. The living machine has, for example, one piece of apparatus capable of being affected by rapidly vibrating waves of air. This bit of the machine we call the ear. It is made of parts delicately adjusted, so that vibrating waves of air set them in motion, and their motion starts a nervous stimulus travelling along the auditory nerve. As a result this apparatus will be set in motion, and an impulse sent along the auditory nerve whenever that external type of motion which we call sound strikes the ear. In other words, the ear is a piece of apparatus for changing air vibrations into nervous stimulation, and is therefore a machine. Apparently the material in the ear is like a bit of gunpowder, capable of being exploded by certain kinds of external excitation; but neither the gunpowder nor the material in the ear develops any energy other than that in it at the outset. In the same way the optic nerve has, at its end, a bit of mechanism readily excited by light vibrations of the ether, and hence the optic nerve will always be excited when ether vibrations chance to have an opportunity of setting the optic machinery in motion. And so on with the other senses. Each sensory nerve has, at its end, a bit of machinery designed for the transformation of certain kinds of external energy into nervous energy, just as a dynamo is a machine for transforming motion into electricity. If the machine is broken, the external force has no longer any power of acting upon it, and the individual becomes deaf or blind.

Mental Phenomena.—Thus far in our analysis we need not hesitate in recognizing a correlation between physical and nervous energy. Even though nervous energy is very subtle and only affects our instruments of measurements under exceptional conditions, the fact that nervous forces are excited by physical forces, and are themselves directly measurable, indicates that they are correlated with physical forces. Up to this point, then, we may confidently say that the nervous system is part of the machine.

But when we turn to the more obscure parts of the nervous phenomena, those which we commonly call mental, we find ourselves obliged to stop abruptly. We may trace the external force to the sensory organ, we may trace this force into a nervous stimulus, and may follow this stimulus to the brain as a wave motion, and therefore as a form of physical energy. But there we must stop. We have no idea of how the nervous impulse is converted into a sensation. The mental side of the sensation appears to stand in a category by itself, and we can not look upon it as a form of energy. It is true that many brave attempts have been made to associate the two. Sensations can be measured as to intensity, and the intensity of a sensation is to a certain extent dependent upon the intensity of the stimulus exciting it. The mental sensation is undoubtedly excited by the physical wave of nervous impulse. In the growth of the individual the development of its mental powers are found to be parallel to the development of its nerves and brain—a fact which, of course, proves that mental power is dependent upon brain structure. Further, it is found that certain visible changes occur in certain parts of the brain—the brain cells—when they are excited into mental activity. Such series of facts point to an association between the mental side of sensations and physical structure of the machine. But they do not prove any correlation between them. The unlikeness of mental and physical phenomena is so absolute that we must hesitate about drawing any connection between them. It is impossible to conceive the mental side of a sensation as a form of wave motion. If, further, we take into consideration the other phenomena associated with the nervous system, the more distinctly mental processes, we have absolutely no data for any comparison. We can not imagine thought measured by units, and until we can conceive of such measurement we can get no meaning from any attempt to find a correlation between mental and physical phenomena. It is true that certain psychologists have tried to build up a conception of the physical nature of mind; but their attempts have chiefly resulted in building up a conception of the physical nature of the brain, and then ignoring the radical chasm that exists between mind and matter. The possibility of describing a complex brain as growing parallel to the growth of a complex mind has been regarded as equivalent to proving their identity. All attempts in this direction thus far have simply ignored the fact that the stimulation of a nerve, a purely physical process, is not the same thing as a mental action. What the future may disclose it is hazardous to say, but at present the mental side of the living machine has not been included within the conception of the mechanical nature of the organism.

The Living Body is a Machine.—Reviewing the subject up to this point, what must be our verdict as to our ability to understand the running of the living machine? In the first place, we are justified in regarding the body as a machine, since, so far as concerns its relations to energy, it is simply a piece of mechanism—complicated, indeed, beyond any other machine, but still a machine for changing one kind of energy into another. It receives the energy in the form of chemical composition and converts it into heat, motion, nervous wave motion, etc. All of this is sure enough. Whether other forms of nervous and mental activity can be placed under the same category, or whether these must be regarded as belonging to a realm by themselves and outside of the scope of energy in the physical sense, can not perhaps be yet definitely decided. We can simply say that as yet no one has been able even to conceive how thought can be commensurate with physical energy. The utter unlikeness of thought and wave motion of any kind leads us at present to feel that on the side of mentality the comparison of the body with a machine fails of being complete.

In regard to the second half of the question, whether natural forces are adequate to explain the running of the machine, we have again been able to reach a satisfactory positive answer. Digestion, assimilation, circulation, respiration, excretion, the principal categories of physiological action, and at least certain phases of the action of the nervous system are readily understood as controlled by the action of chemical and physical forces. In the accomplishment of these actions there is no need for the supposition of any force other than those which are at our command in the scientific laboratory.

The Living Machine Constructive as well as Destructive.—In one respect the living machine differs from all others. The action of all other machines results in the destruction of organized material, and thus in a degradation of matter. For example, a steam engine receives coal, a substance of high chemical composition, and breaks it into more simple compounds, in this way liberating its stored energy. Now if we examine all forms of artificial machines, we find in the same way that there is always a destruction of compounds of high chemical composition. In such machines it is common to start with heat as a source of energy, and this heat is always produced by the breaking of chemical compounds to pieces. In all chemical processes going on in the chemist's laboratory there is similarly a destruction of organic compounds. It is true that the chemist sometimes makes complex compounds out of simpler ones; but in order to do this he is obliged to use heat to bring about the combination, and this heat is obtained from the destruction of a much larger quantity of high compounds than he manufactures. The total result is therefore destruction rather than manufacture of high compounds. Thus it is a fact, that in all artificial machines and in all artificial chemical processes there is, as a total result, a degradation of matter toward the simpler from the more complex compounds.

As a result of the action of the living machine, however, we have the opposite process of construction going on. All high chemical compounds are to be traced to living beings as their source. When green plants grow in sunlight they take simple compounds and combine them together to form more complex ones in such a way that the total result is an increase of chemical compounds of high complexity. In doing this they use the energy of sunlight, which they then store away in the compounds formed. They thus produce starches, oils, proteids, woods, etc., and these stores of energy now may be used by artificial machines. The living machine builds up, other machines pull down. The living machine stores sunlight in complex compounds, other machines take it out and use it. The living organism is therefore to be compared to a sun engine, which obtains its energy directly from the sun, rather than to the ordinary engine. While this does not in the slightest militate against the idea of the living body as a machine, it does indicate that it is a machine of quite a different character from any other, and has powers possessed by no other machine. Living machines alone increase the amount of chemical compounds of high complexity.

We must notice, however, that this power of construction in distinction from destruction, is possessed only by one special class of living machines. Green plants alone can thus increase the store of organic compounds in the world. All colourless plants and all animals, on the other hand, live by destroying these compounds and using the energy thus liberated; in this respect being more like ordinary artificial machines. The animal does indeed perform certain constructive operations, manufacturing complex material out of simpler bodies; as, for example, making fats out of starches. But in this operation it destroys a large amount of organic material to furnish the energy for the construction, so that the total result is a degradation of chemical compounds rather than a construction. Constructive processes, which increase the amount of high compounds in nature, are confined to the living machine, and indeed to one special form of it, viz., the green plant. This constructive power radically separates the living from other machines; for while constructive processes are possible to the chemist, and while engines making use of sunlight are possible, the living machine is the only machine that increases the amount of high chemical compounds in the world.

The Vital Factor.—With all this explanation of life processes it can not fail to be apparent that we have not really reached the centre of the problem. We have explained many secondary processes, but the primary ones are still unsolved. In studying digestion we reach an understanding of everything until we come to the active vital property of the gland-cells in secreting. In studying absorption we understand the process until we come to what we have called the vital powers of the absorptive cells of the alimentary canal. The circulation is intelligible until we come to the beating of the heart and the contraction of the muscles of the blood-vessels. Excretion is also partly explained, but here again we finally must refer certain processes to the vital powers of active cells. And thus wherever we probe the problem we find ourselves able to explain many secondary problems, while the fundamental ones we still attribute to the vital properties of the active tissues. Why a muscle contracts or a gland secretes we have certainly not yet answered. The relation of the actions to the general problems of correlation of force is simple enough. That a muscle is a machine in the sense of our definition is beyond question. But the problem of why a muscle acts is not answered by showing that it derives its energy from broken food material. There are plainly still left for us a number of fundamental problems, although the secondary ones are soluble.

What can we say in regard to these fundamental vital powers of the active tissues? Firstly, we must notice that many of the processes which we now understand were formerly classed as vital, and we only retain under this term those which are not yet explained. This, of course, suggests to us that perhaps we may some day find an explanation for all the so-called vital powers by the application of simple physical forces. Is it a fact that the only significance to the term vital is that we have not yet been able to explain these processes to our entire satisfaction? Is the difference between what we have called the secondary processes and the primary ones only one of degree? Is there a probability that the actions which we now call vital will some day be as readily understood as those which have already been explained?

Is there any method by which we can approach these fundamental problems of muscle action, heart beat, gland secretion, etc.? Evidently, if this is to be done, it must be by resolving the body into its simple units and studying these units. Our study thus far has been a study of the machinery of the body as a whole; but we have found that the various parts of the machine are themselves active, that apart from the action of the general machine as a whole, the separate parts have vital powers. We must, therefore, get rid of this complicated machinery, which confuses the problem, and see if we can find the fundamental units which show these properties, unencumbered by the secondary machinery which has hitherto attracted our attention. We must turn now to the problem connected with protoplasm and the living cell, since here, if anywhere, can we find the life substance reduced to its lowest terms.

CHAPTER II.
THE CELL AND PROTOPLASM.

Vital Properties.—We have seen that the general activities of the body are intelligible according to chemical and mechanical laws, provided we can assume as their foundation the simple vital properties of living phenomena. We must now approach closer to the centre of the problem, and ask whether we can trace these fundamental properties to their source and find an explanation of them.

In the first place, what are these properties? The vital powers are varied, and lie at the basis of every form of living activity. When we free them from complications, however, they may all be reduced to four. These are: (1) Irritability, or the property possessed by living matter of reacting when stimulated. (2) Movement, or the power of contracting when stimulated. (3) Metabolism, or the power of absorbing extraneous food and producing in it certain chemical changes, which either convert it into more living tissue or break it to pieces to liberate the inclosed energy. (4) Reproduction, or the power of producing new individuals. From these four simple vital activities all other vital actions follow; and if we can find an explanation of these, we have explained the living machine. If we grant that certain parts of the body can assimilate food and multiply, having the power of contraction when irritated, we can readily explain the other functions of the living machine by the application of these properties to the complicated machinery of the body. But these properties are fundamental, and unless we can grasp them we have failed to reach the centre of the problem.

As we pass from the more to the less complicated animals we find a gradual simplification of the machinery until the machinery apparently disappears. With this simplification of the machinery we find the animals provided with less varied powers and with less delicate adaptations to conditions. But withal we find the fundamental powers of the living organisms the same. For the performance of these fundamental activities there is apparently needed no machinery. The simple types of living bodies are simple in number of parts, but they possess essentially the same powers of assimilation and growth that characterize the higher forms. It is evident that in our attempt to trace the vital properties to their source we may proceed in two ways. We may either direct our attention to the simplest organisms where all secondary machinery is wanting, or to the smallest parts into which the tissues of higher organisms can be resolved and yet retain their life properties. In either way we may hope to find living phenomena in its simplest form independent of secondary machinery.

But the fact is, when we turn our attention in these two directions, we find the result is the same. If we look for the lowest organisms we find them among forms that are made of a single cell, and if we analyze the tissues of higher animals we find the ultimate parts to be cells. Thus, in either direction, the study of the cell is forced upon us.

Before beginning the study of the cell it will be well for us to try to get a clear notion of the exact nature of the problems we are trying to solve. We wish to explain the activities of life phenomena in such a way as to make them intelligible through the application of natural forces. That these processes are fundamentally chemical ones is evident enough. A chemical oxidation of food lies at the basis of all vital activity, and it is thus through the action of chemical forces that the vital powers are furnished with their energy. But the real problem is what it is in the living machine that controls these chemical processes. Fat and starch may be oxidized in a chemist's test tubes, and will there liberate energy; but they do not, under these conditions, manifest vital phenomena. Proteid may be brought in contact with oxygen without any oxidation occurring, and even if it is oxidized no motion or assimilation or reproduction occurs under ordinary conditions. These phenomena occur only when the oxidation takes place in the living machine. Our problem is then to determine, if possible, what it is in the living machine that regulates the oxidations and other changes in such a way as to produce from them vital activities. Why is it that the oxidation of starch in the living machine gives rise to motion, growth, and reproduction, while if the oxidation occurs in the chemist's laboratory, or even in a bit of dead protoplasm, it simply gives rise to heat?

One of the primary questions to demand attention in this search is whether we are to find the explanation, at the bottom, a chemical or a mechanical one. In the simplest form of life in which vital manifestations are found are we to attribute these properties simply to chemical forces of the living substance, or must we here too attribute them to the action of a complicated machinery? This question is more than a formal one. That it is one of most profound significance will appear from the following considerations:

Chemical affinity is a well recognized force. Under the action of this force chemical compounds are produced and different compounds formed under different conditions. The properties of the different compounds differ with their composition, and the more complex are the compounds the more varied their properties. Now it might be assumed as an hypothesis that there could be a chemical compound so complex as to possess, among other properties, that of causing the oxidation of food to occur in such a way as to produce assimilation and growth. Such a compound would, of course, be alive, and it would be just as true that its power of assimilating food would be one of its physical properties as it is that freezing is a physical property of water. If such an hypothesis should prove to be the true one, then the problem of explaining life would be a chemical one, for all vital properties would be reducible to the properties of a chemical compound. It would then only be necessary to show how such a compound came into existence and we should have explained life. Nor would this be a hopeless task. We are well acquainted with forces adequate to the formation of chemical compounds. If the force of chemical affinity is adequate under certain conditions to form some compounds, it is easy to conceive it as a possibility under other conditions to produce this chemical living substance. Our search would need then to be for a set of conditions under which our living compound could have been produced by the known forces of chemical affinity.

But suppose, on the other hand, that we find this simplest bit of living matter is not a chemical compound, but is in itself a complicated machine. Suppose that, after reducing this vital substance to its simplest type, we find that the substance with which we are dealing not only has complex chemical structure, but that it also possesses a large number of structural parts adapted to each other in such a way as to work together in the form of an intricate mechanism. The whole problem would then be changed. To explain such a machine we could no longer call upon chemical forces. Chemical affinity is adequate to the explanation of chemical compounds however complicated, but it cannot offer any explanation for the adaptation of parts which make a machine. The problem of the origin of the simplest form of life would then be no longer one of chemical but one of mechanical evolution. It is plain then that the question of whether we can attribute the properties of the simplest type of life to chemical composition or to mechanical structure is more than a formal one.

The Discovery of Cells.—It is difficult for us to-day to have any adequate idea of the wonderful flood of light that was thrown upon scientific and philosophical study by the discoveries which are grouped around the terms cells and protoplasm. Cells and protoplasm have become so thoroughly a part of modern biology that we can hardly picture to ourselves the vagueness of knowledge before these facts were recognized. Perhaps a somewhat crude comparison will illustrate the relation which the discovery of cells had to the study of life.

Imagine for a moment, some intelligent being located on the moon and trying to study the phenomena on the earth's surface. Suppose that he is provided with a telescope sufficiently powerful to disclose moderately large objects on the earth, but not smaller ones. He would see cities in various parts of the world with wide differences in appearance, size, and shape. He would see railroad trains on the earth rushing to and fro. He would see new cities arising and old ones increasing in size, and we may imagine him speculating as to their method of origin and the reasons why they adopt this or that shape. But in spite of his most acute observations and his most ingenious speculation, he could never understand the real significance of the cities, since he is not acquainted with the actual living unit. Imagine now, if you will, that this supramundane observer invents a telescope which enables him to perceive more minute objects and thus discovers human beings. What a complete revolution this would make in his knowledge of mundane affairs! We can imagine how rapidly discovery would follow discovery; how it would be found that it was the human beings that build the houses, construct and run the railroads, and control the growth of the cities according to their fancy; and, lastly, how it would be learned that it is the human being alone that grows and multiplies and that all else is the result of his activities. Such a supramundane observer would find himself entering into a new era, in which all his previous knowledge would sink into oblivion.

Something of this same sort of revolution was inaugurated in the study of living things by the discovery of cells and protoplasms. Animals and plants had been studied for centuries and many accurate and painstaking observations had been made upon them. Monumental masses of evidence had been collected bearing upon their shapes, sizes, distribution, and relations. Anatomy had long occupied the attention of naturalists, and the general structure of animals and plants was already well known. But the discoveries starting in the fourth decade of the century by disclosing the unity of activity changed the aspect of biological science.

The Cell Doctrine.—The cell doctrine is, in brief, the theory that the bodies of animals and plants are built up entirely of minute elementary units, more or less independent of each other, and all capable of growth and multiplication. This doctrine is commonly regarded as being inaugurated in 1839 by Schwann. Long before this, however, many microscopists had seen that the bodies of plants are made up of elementary units. In describing the bark of a tree in 1665, Robert Hooke had stated that it was composed of little boxes or cells, and regarded it as a sort of honeycomb structure with its cells filled with air. The term cell quite aptly describes the compartments of such a structure, as can be seen by a glance at Fig. 7, and this term has been retained even till to-day in spite of the fact that its original significance has entirely disappeared. During the last century not a few naturalists observed and described these little vesicles, always regarding them as little spaces and never looking upon them as having any significance in the activities of plants. In one or two instances similar bodies were noticed in animals, although no connection was drawn between them and the cells of plants. In the early part of the century observations upon various kinds of animals and plant tissues multiplied, and many microscopists independently announced the discovery of similar small corpuscular bodies. Finally, in 1839, these observations were combined together by Schwann into one general theory. According to the cell doctrine then formulated, the parts of all animals and plants are either composed of cells or of material derived from cells. The bark, the wood, the roots, the leaves of plants are all composed of little vesicles similar to those already described under the name of cells. In animals the cellular structure is not so easy to make out; but here too the muscle, the bone, the nerve, the gland are all made up of similar vesicles or of material made from them. The cells are of wonderfully different shapes and widely different sizes, but in general structure they are alike. These cells, thus found in animals and plants alike, formed the first connecting link between animals and plants. This discovery was like that of our supposed supramundane observer when he first found the human being that brought into connection the widely different cities in the various parts of the world.

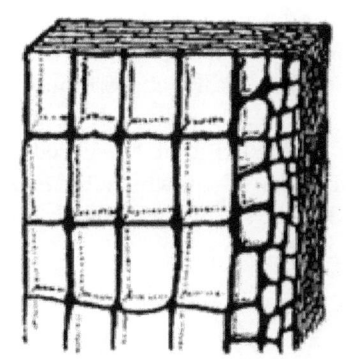

FIG. 7.—A bit of bark showing cellular structure.

Schwann and his immediate followers, while recognizing that the bodies of animals and plants were composed of cells, were at a loss to explain how these cells arose. The belief held at first was that there existed in the bodies of animals and plants a structureless substance which formed the basis out of which the cells develop, in somewhat the same way that crystals arise from a mother liquid. This supposed substance Schwann called the cytoblastema, and he thought it existed between the cells or sometimes within them. For example, the fluid part of the blood is the cytoblastema, the blood corpuscles being the cells. From this structureless fluid the cells were supposed to arise by a process akin to crystallization. To be sure, the cells grow in a manner very different from that of a crystal. A crystal always grows by layers being added upon its outside, while the cells grow by additions within its body. But this was a minor detail, the essential point being that from a structureless liquid containing proper materials the organized cell separated itself.

This idea of the cytoblastema was early thrown into suspicion, and almost at the time of the announcement of the cell doctrine certain microscopists made the claim that these cells did not come from any structureless medium, but by division from other cells like themselves. This claim, and its demonstration, was of even greater importance than the discovery of the cells. For a number of years, however, the matter was in dispute, evidence being collected which about equally attested each view. It was a Scotchman, Dr. Barry, who finally produced evidence which settled the question from the study of the developing egg.

The essence of his discovery was as follows: The ovum of an animal is a single cell, and when it begins to develop into an embryo it first simply divides into two halves, producing two cells (Fig, 8, a and b). Each of these in turn divides, giving four, and by repeated divisions of this kind there arises a solid mass of smaller cells (Fig. 8, b to f,) called the mulberry stage, from its resemblance to a berry. This is, of course, simply a mass of cells, each derived by division from the original. As the cells increase in number, the mass also increases in size by the absorption of nutriment, and the cells continue dividing until the mass contains thousands of cells. Meantime the body of the animal is formed out of these cells, and when it is adult it consists of millions of cells, all of which have been derived by division from the original cell. In such a history each cell comes from pre-existing cells and a cytoblastema plays no part.

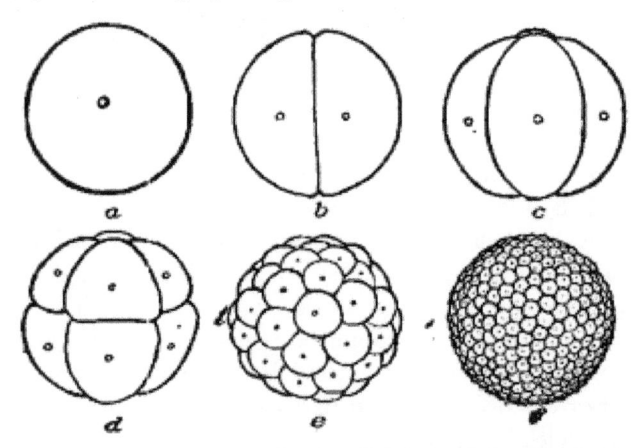

FIG. 8.—Successive stages in the division of the developing egg.

It was impossible, however, for Barry or any other person to follow the successive divisions of the egg cell through all the stages to the adult. The divisions can be followed for a short time under the microscope, but the rest must be a matter of simple inference. It was argued that since cell origin begins in this way by simple division, and since the same process can be observed in the adult, it is reasonable to assume that the same process has continued uninterruptedly, and that this is the only method of cell origin. But a final demonstration of this conclusion was not forthcoming for a long time. For many years some biologists continued to believe that cells can have other origin than from pre-existing cells. Year by year has the evidence for such "free cell" origin become less, until the view has been entirely abandoned, and to-day it is everywhere admitted that new cells always arise from old ones by direct descent, and thus every cell in the body of an animal or plant is a direct descendant by division from the original egg cell.

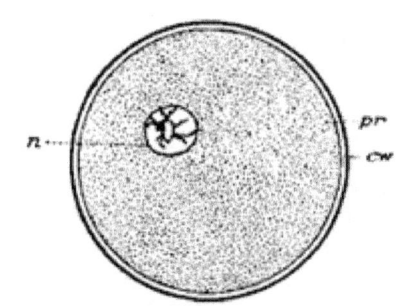

FIG. 9.—A cell; cw is the cell wall; pr, the cell substance; n, the nucleus.

The Cell.—But what is this cell which forms the unit of life, and to which all the fundamental vital properties can be traced? We will first glance at the structure of the cell as it was understood by the earlier microscopists. A typical cell is shown in Fig. 9. It will be seen that it consists of three quite distinct parts. There is first the cell wall (cw) which is a limiting membrane of varying thickness and shape. This is in reality lifeless material, and is secreted by the rest of the cell. Being thus produced by the other active parts of the cell, we will speak of it as formed material in distinction from the rest, which is active material. Inside this vesicle is contained a somewhat transparent semifluid material which has received various names, but which for the present we will call cell substance (Fig. 9, pr). It may be abundant or scanty, and has a widely varying consistency from a very liquid mass to a decidedly thick jellylike substance. Lying within the cell substance is a small body, usually more or less spherical in shape, which is called the nucleus (Fig. 9, n). It appears to the microscope similar to the cell substance in character, and has frequently been described as a bit of the cell substance more dense than the remainder. Lying within the nucleus there are usually to be seen one or more smaller rounded bodies which have been called nucleoli. From the very earliest period that cells have been studied, these three parts, cell wall, cell substance, and nucleus have been recognized, but as to their relations to each other and to the general activities of the cell there has been the widest variety of opinion.

Cellular Structure of Organisms.—It will be well to notice next just what is meant by saying that all living bodies are composed of cells. This can best be understood by referring to the accompanying figures. Figs. 10-14, for instance, show the microscopic appearance of several plant tissues.

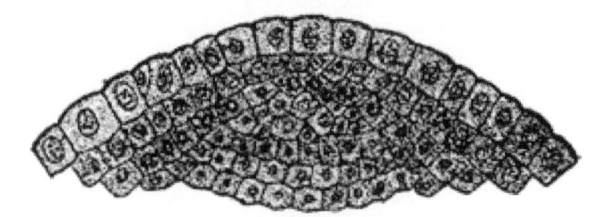

FIG. 10.—Cells at a root tip.

At Fig. 10 will be seen the tip of a root, plainly made of cells quite similar to the typical cell described. At Fig. 11 will be seen a bit of a leaf showing the same general structure. At Fig. 12 is a bit of plant tissue of which the cell walls are very thick, so that a very dense structure is formed.

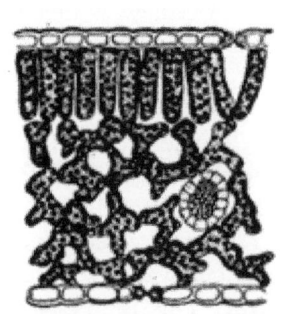

FIG. 11.—Section of a leaf showing cells of different shapes.

At Fig. 13 is a bit of a potato showing its cells filled with small granules of starch which the cells have produced by their activities and deposited within their own bodies. At Fig. 14 are several wood cells showing cell walls of different shape which, having become dead, have lost their contents and simply remain as dead cell walls. Each was in its earlier history filled with cell substance and contained a nucleus. In a similar way any bit of vegetable tissue would readily show itself to be made of similar cells.

In animal tissues the cellular structure is not so easily seen, largely because the products made by the cells, the formed products, become relatively more abundant and the cells themselves not so prominent. But the cellular structure is none the less demonstrable. In Fig. 15, for instance, will be seen a bit of cartilage where the cells themselves are rather small, while the material deposited between them is abundant. This material between the cells is really to be regarded as an excessively thickened cell wall and has been secreted by the cell substance lying within the cells, so that a bit of cartilage is really a mass of cells with an exceptionally thick cell wall.

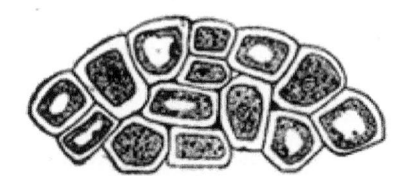

FIG. 12.—Plant cells with thick walls, from a fern.

At Fig. 16 is shown a little blood. Here the cells are to be seen floating in a liquid. The liquid is colourless and it is the red colour in the blood cells which gives the blood its red colour. The liquid may here again be regarded as material produced by cells. At Fig. 17 is a bit of bone showing small irregular cells imbedded within a large mass of material which has been deposited by the cell.

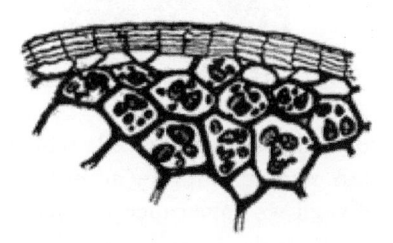

FIG. 13.—Section of a potato showing different shaped cells, the inner and larger ones being filled with grains of starch.

In this case the formed material has been hardened by calcium phosphate, which gives the rigid consistency to the bone. In some animal tissues the formed material is still greater in amount. At Fig. 18, for example, is a bit of connective tissue, made up of a mass of fine fibres which have no resemblance to cells, and indeed are not cells. These fibres have, however, been made by cells, and a careful study of such tissue at proper places will show the cells within it. The cells shown in Fig. 18 (c) have secreted the fibrous material. Fig. 19 shows a cell composing a bit of nerve. At Fig. 20 is a bit of muscle; the only trace of cellular structure that it shows is in the nuclei (n), but if the muscle be studied in a young condition its cellular structure is more evident.

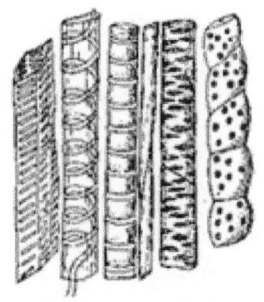

FIG. 14.—Various shaped wood cells from plant tissue.

FIG. 15.—A bit of cartilage.

Thus it happens in adult animals that the cells which are large and clear at first, become less and less evident, until the adult tissue seems sometimes to be composed mostly of what we have called formed material.

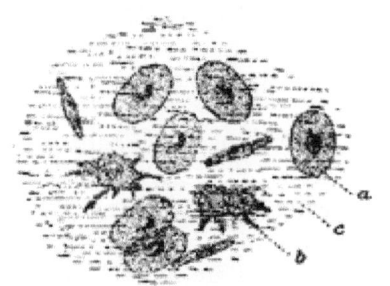

FIG. 16.—Frog's blood: a and b are the cells; c is the liquid.

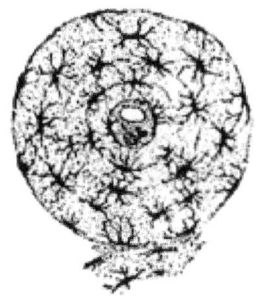

FIG. 17.—A bit of bone, showing
the cells imbedded in the bony matter.

It must not be imagined, however, that a very rigid line can be drawn between the cell itself and the material it forms. The formed material is in many cases simply a thickened cell wall, and this we commonly regard as part of the cell. In many cases the formed material is simply the old dead cell walls from which the living substance has been withdrawn (Fig. 14). In other cases the cell substance acquires peculiar functions, so that what seems to be the formed material is really a modified cell body and is still active and alive. Such is the case in the muscle. In other cases the formed material appears to be manufactured within the cell and secreted, as in the case of bone. No sharp lines can be drawn, however, between the various types. But the distinction between formed material and cell body is a convenient one and may well be retained in the discussion of cells. In our discussion of the fundamental vital properties we are only concerned in the cell substance, the formed material having nothing to do with fundamental activities of life, although it forms largely the secondary machinery which we have already studied.

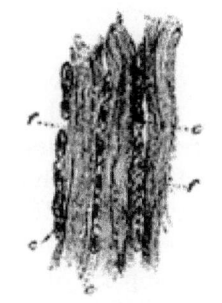

FIG. 18.—Connective
tissue. The cells
of the tissue are
shown at c, and the

fibres or formed
matter at f.

In all higher animals and plants the life of the individual begins as a single ovum or a single cell, and as it grows the cells increase rapidly until the adult is formed out of hundreds of millions of cells. As these cells become numerous they cease, after a little, to be alike. They assume different shapes which are adapted to the different duties they are to perform. Thus, those cells which are to form bone soon become different from those which are to form muscle, and those which are to form the blood are quite unlike those which are to produce the hairs. By means of such a differentiation there arises a very complex mass of cells, with great variety in shape and function.

FIG. 19. A piece of nerve fibre, showing the cell with its nucleus at n.

It should be noticed further that there are some animals and plants in which the whole animal is composed of a single cell. These organisms are usually of extremely minute size, and they comprise most of the so-called animalculæ which are found in water. In such animals the different parts of the cell are modified to perform different functions. The different organs appear within the cell, and the cell is more complex than the typical cell described. Fig. 21 shows such a cell. Such an animal possesses several organs, but, since it consists of a single mass of protoplasm and a single nucleus, it is still only a single cell. In the multicellular organisms the organs of the body are made up of cells, and the different organs are produced by a differentiation of cells, but in the unicellular organisms the organs are the results of the differentiation of the parts of a single cell. In the one case there is a differentiation of cells, and in the other of the parts of a cell.

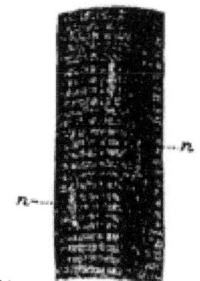

FIG. 20.—A muscle fibre. The nucleii are shown at n.

Such, in brief, is the cell to whose activities it is possible to trace the fundamental properties of all living things. Cells are endowed with the properties of irritability, contractibility, assimilation and reproduction, and it is thus plainly to the study of cells that we must look for an interpretation of life phenomena. If we can reach an intelligible understanding of the activities of the cell our problem is solved, for the activities of the fully formed animal or plant, however complex, are simply the application of mechanical and chemical principles among the groups of such cells. But wherein does this knowledge of cells help us? Are we any nearer to understanding how these vital processes arise? In answer to this question we may first ask whether it is possible to determine whether any one part of the cell is the seat of its activities.

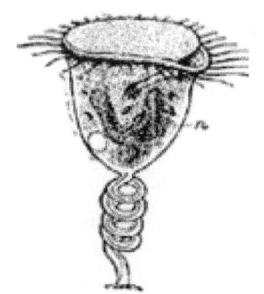

FIG. 21.—A complex cell. It is an entire animal, but composed of only one cell.

The Cell Wall.—The first suggestion which arose was that the cell wall was the important part of the cell, the others being secondary. This was not an unnatural conclusion. The cell wall is the most persistent part of the cell. It was the part first discovered by the microscope and is the part which remains after the other parts are gone. Indeed, in many of the so-called cells the cell wall is all that is seen, the cell contents having disappeared (Fig. 14). It was not strange, then, that this should at first have been looked upon as the primary part. The idea was that the cell wall in some way changed the chemical character of the substances in contact with its two sides, and thus gave rise to vital activities which, as we have seen, are fundamentally chemical. Thus the cell wall was regarded as the most essential part of the cell, since it controlled its activities. This the belief of Schwann, although he also regarded the other parts of the cell as of importance.

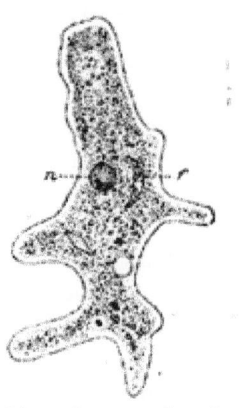

FIG. 22.—An amœba. A single cell without cell wall. n is the nucleus; f, a bit of food which the cell has absorbed.

This conception, however, was quite temporary. It was much as if our hypothetical supramundane observer looked upon the clothes of his newly discovered human being as forming the essential part of his nature. It was soon evident that this position could not be maintained. It was found that many bits of living matter were entirely destitute of cell wall. This is especially true of animal cells. While among plants the cell wall is almost always well developed, it is very common for animal cells to be entirely lacking in this external covering—as, for example, the white blood-cells. Fig. 22 shows an amœba, a cell with very active powers of motion and assimilation, but with no cell wall. Moreover, young cells are always more active than older ones, and they commonly possess either no cell wall or a very slight one, this being deposited as the cell becomes older and remaining long after it is dead. Such facts soon disproved the notion that the cell wall is a vital part of the cell, and a new conception took its place which was to have a more profound influence upon the study of living things than any discovery hitherto made. This was the formulation of the doctrine of the nature of protoplasm.

Protoplasm.—(a) Discovery. As it became evident that the cell wall is a somewhat inactive part of the cell, more attention was put on the cell contents. For twenty years after the formulation of the cell doctrine both the cell substance and the nucleus had been looked upon as essential to its activities. This was more especially true of the nucleus, which had been thought of as an organ of reproduction. These suggestions appeared indefinitely in the writings of one scientist and another, and were finally formulated in 1860 into a general theory which formed what has sometimes been called the starting point of modern biology. From that time the material known as protoplasm was elevated into a prominent position in the discussion of all subjects connected with living phenomena. The idea of protoplasm was first clearly defined by Schultze, who claimed that the real active part of the cell was the cell substance within the cell wall. This substance he proved to be endowed with powers of motion and powers of inducing chemical changes associated with vital phenomena. He showed it to be the most abundant in the most active cells, becoming less abundant as the cells lose their activity, and disappearing when the cells lose their vitality. This cell substance was soon raised into a position of such importance that the smaller body within it was obscured, and for some twenty years more the nucleus was silently ignored in biological discussion. According to Schultze, the cell substance itself constituted the cell, the other parts being entirely subordinate, and indeed frequently absent. A cell was thus a bit of protoplasm, and nothing more. But the more important feature of this doctrine was not the simple conclusion that the cell substance constitutes the cell, but the more sweeping conclusion that this cell substance is in all cells essentially identical. The study of all animals, high and low, showed all active cells filled with a similar material, and more important still, the study of plant cells disclosed a material strikingly similar. Schultze experimented with this material by all means at his command, and finding that the cell substance in all animals and plants obeys the same tests, reached the conclusion that the cell substance in animals and plants is always identical. To this material he now gave the name protoplasm, choosing a name hitherto given to the cell contents of plant cells. From this time forth this term protoplasm was applied to the living material found in all cells, and became at once the most important factor in the discussion of biological problems.

The importance of this newly formulated doctrine it is difficult to appreciate. Here, in protoplasm had been apparently found the foundation of living phenomena. Here was a substance universally present in animals and plants, simple and uniform—a substance always present in living parts and disappearing with death. It was the simplest thing that had life, and indeed the only thing that had life, for there is no life outside of cells and protoplasm. But simple as it was it had all the fundamental properties of living things—irritability, contractibility, assimilation, and reproduction. It was a compound which seemingly deserved the name of "physical basis of life", which was soon given to it by Huxley. With this conception of protoplasm as the physical basis of life the problems connected with the study of life became more simplified. In order to study the nature of life it was no longer necessary to study the confusing mass of complex organs disclosed to us by animals and plants, or even the somewhat less confusing structures shown by individual cells. Even the simple cell has several separate parts capable of undergoing great modifications in different types of animals. This confusion now appeared to vanish, for only one thing was found to be alive, and that was apparently very simple. But that substance exhibited all the properties of life. It moved, it could grow, and reproduce itself, so that it was necessary only to explain this substance and life would be explained.

(b) Nature of Protoplasm.—What is this material, protoplasm? As disclosed by the early microscope it appeared to be nothing more than a simple mass of jelly, usually transparent, more or less consistent, sometimes being quite fluid, and at others more solid. Structure it appeared to have none. Its chief peculiarity, so far as physical characters were concerned, was a wonderful and never-ceasing activity. This jellylike material appeared to be endowed with wonderful powers, and yet neither physical nor microscopical study revealed at first anything more than a uniform homogeneous mass of jelly. Chemical study of the same substance was of no less interest than the microscopical study. Of course it was no easy matter to collect this protoplasm in sufficient quantity and pure enough to make a careful analysis. The difficulties were in time, however, overcome, and chemical study showed protoplasm to be a proteid, related to other proteids like albumen, but one which was more complex than any other known. It was for a long time looked upon by many as a single definite chemical compound, and attempts were made to determine its chemical formula. Such an analysis indicated a molecule made up of several hundred atoms. Chemists did not, however, look with much confidence upon these results, and it is not surprising that there was no very close agreement among them as to the number of atoms in this supposed complex molecule. Moreover, from the very first, some biologists thought protoplasm to be not one, but more likely a mixture of several substances. But although it was more complex than any other substance studied, its general characters were so like those of albumen that it was uniformly regarded as a proteid; but one which was of a higher complexity than others, forming perhaps the highest number of a series of complex chemical compounds, of which ordinary proteids, such as albumen, formed lower members. Thus, within a few years following the discovery of protoplasm there had developed a theory that living phenomena are due to the activities of a definite though complex chemical compound, composed chiefly of the elements carbon, oxygen, hydrogen, and nitrogen, and closely related to ordinary proteids. This substance was the basis of living activity, and to its modification under different conditions were due the miscellaneous phenomena of life.

(c) Significance of Protoplasm.—The philosophical significance of this conception was very far-reaching. The problem of life was so simplified by substituting the simple protoplasm for the complex organism that its solution seemed to be not very difficult. This idea of a chemical compound as the basis of all living phenomena gave rise in a short time to a chemical theory of life which was at least tenable, and which accounted for the fundamental properties of life. That theory, the chemical theory of life, may be outlined somewhat as follows:

The study of the chemical nature of substances derived from living organisms has developed into what has been called organic chemistry. Organic chemistry has shown that it is possible to manufacture artificially many of the compounds which are called organic, and which had been hitherto regarded as produced only by living organisms. At the beginning of the century, it was supposed to be impossible to manufacture by artificial means any of the compounds which animals and plants produce as the result of their life. But chemists were not long in showing that this position is untenable. Many of the organic products were soon shown capable of production by artificial means in the chemist's laboratory. These organic compounds form a series beginning with such simple bodies as carbonic acid (CO_2), water (H_2O), and ammonia (NH_3), and passing up through a large number of members of greater and greater complexity, all composed, however, chiefly of the elements carbon, oxygen, hydrogen, and nitrogen. Our chemists found that starting with simple substances they could, by proper means, combine them into molecules of greater complexity, and in so doing could make many of the compounds that had hitherto been produced only as a result of living activities. For example, urea, formic acid, indigo, and many other bodies, hitherto produced only by animals and plants, were easily produced by the chemist by purely chemical methods. Now when protoplasm had been discovered as the "physical basis of life," and, when it was further conceived that this substance is a proteid related to albumens, it was inevitable that a theory should arise which found the explanation of life in accordance with simple chemical laws.

If, as chemists and biologists then believe, protoplasm is a compound which stands at the head of the organic series, and if, as is the fact, chemists are each year succeeding in making higher and higher members of the series, it is an easy assumption that some day they will be able to make the highest member of the series. Further, it is a well-known fact that simple chemical compounds have simple physical properties, while the higher ones have more varied properties. Water has the property of being liquid at certain temperatures and solid at others, and of dividing into small particles (i.e., dissolving) certain bodies brought in contact with it. The higher compound albumen has, however, a great number of properties and possibilities of combination far beyond those of water. Now if the properties increase in complexity with the complexity of the compound, it is again an easy assumption that when we reach a compound as complex as protoplasm, it will have properties as complex as those of the simple life substance. Nor was this such a very wild hypothesis. After all, the fundamental life activities may all be traced to the simple oxidation of food, for this results in movement, assimilation, and growth, and the result of growth is reproduction. It was therefore only necessary for our biological chemists to suppose that their chemical compound protoplasm possessed the power of causing certain kinds of oxidation to take place, just as water itself induces a simpler kind of oxidation, and they would have a mechanical explanation of the life activities. It was certainly not a very absurd assumption to make, that this substance protoplasm could have this power, and from this the other vital activities are easily derived.

In other words, the formulation of the doctrine of protoplasm made it possible to assume that life is not a distinct force, but simply a name given to the properties possessed by that highly complex chemical compound protoplasm. Just as we might give the name aquacity to the properties possessed by water, so we have actually given the name vitality to the properties possessed by protoplasm. To be sure, vitality is more marvelous than aquacity, but so is protoplasm a more complex compound than water. This compound was a very unstable compound, just as is a mass of gunpowder, and hence it is highly irritable, also like gunpowder, and any disturbance of its condition produces motion, just as a spark will do in a mass of gunpowder. It is capable of inducing oxidation in foods, something as water induces oxidation in a bit of iron. The oxidation is, however, of a different kind, and results in the formation of different chemical combinations; but it is the basis of assimilation. Since now assimilation is the foundation of growth and reproduction, this mechanical theory of life thus succeeded in tracing to the simple properties of the chemical compound protoplasm, all the fundamental properties of life. Since further, as we have seen in our first chapter, the more complex properties of higher organisms are easily deduced from these simple ones by the application of the laws of mechanics, we have here in this mechanical theory of life the complete reduction of the body to a machine.

The Reign of Protoplasm.—This substance protoplasm became now naturally the centre of biological thought. The theory of protoplasm arose at about the same time that the doctrine of evolution began to be seriously discussed under the stimulus of Darwin, and naturally these two great conceptions developed side by side. Evolution was constantly teaching that natural forces are sufficient to account for many of the complex phenomena which had hitherto been regarded as insolvable; and what more natural than the same kind of thinking should be applied to the vital activities manifested by this substance protoplasm. While the study of plants and animals was showing scientists that natural forces would explain the origin of more complex types from simpler ones through the law of natural selection, here in this conception of protoplasm was a theory which promised to show how the simplest forms may have been derived from the non-living. For an explanation of the origin of life by natural means appeared now to be a simple matter.

It required now no violent stretch of the imagination to explain the origin of life something as follows: We know that the chemical elements have certain affinities for each other, and will unite with each other under proper conditions. We know that the methods of union and the resulting compounds vary with the conditions under which the union takes place. We know further that the elements carbon, hydrogen, oxygen, and nitrogen have most remarkable properties, and unite to form an almost endless series of remarkable bodies when brought into combination under different conditions. We know that by varying the conditions the chemist can force these elements to unite into a most extraordinary variety of compounds with an equal variety of properties. What more natural, then, than the assumption that under certain conditions these same elements would unite in such a way as to form this compound protoplasm; and then, if the ideas concerning protoplasm were correct, this body would show the properties of protoplasm, and therefore be alive. Certainly such a supposition was not absurd, and viewed in the light of the rapid advance in the manufacture of organic compounds could hardly be called improbable. Chemists beginning with simple bodies like CO_2 and H_2O were climbing the ladder, each round of which was represented by compounds of higher complexity. At the top was protoplasm, and each year saw our chemists nearer the top of the ladder, and thus approaching protoplasm as their final goal. They now began to predict that only a few more years would be required for chemists to discover the proper conditions, and thus make protoplasm. As late as 1880 the prediction was freely made that the next great discovery would be the manufacture of a bit of protoplasm by artificial means, and thus in the artificial production of life. The rapid advance in organic chemistry rendered this prediction each year more and more probable. The ability of chemists to manufacture chemical compounds appeared to be unlimited, and the only question in regard to their ability to make protoplasm thus resolved itself into the question of whether protoplasm is really a chemical compound.

We can easily understand how eager biologists became now in pursuit of the goal which seemed almost within their reach; how interested they were in any new discovery, and how eagerly they sought for lower and simpler types of protoplasm since these would be a step nearer to the earliest undifferentiated life substance. Indeed so eager was this pursuit for pure undifferentiated protoplasm, that it led to one of those unfounded discoveries which time showed to be purely imaginary. When this reign of protoplasm was at its height and biologists were seeking for even greater simplicity a most astounding discovery was announced. The British exploring ship Challenger had returned from its voyage of discovery and collection, and its various treasures were turned over to the different scientists for study. The brilliant Prof. Huxley, who had first formulated the mechanical theory of life, now startled the biological world with the statement that these collections had shown him that at the bottom of the deep sea, in certain parts of the world, there exists a diffused mass of living undifferentiated protoplasm. So simple and undifferentiated was it that it was not divided into cells and contained no nucleii. It was, in short, exactly the kind of primitive protoplasm which the evolutionist wanted to complete his chain of living structures, and the biologist wanted to serve as a foundation for his mechanical theory of life. If such a diffused mass of undifferentiated protoplasm existed at the bottom of the sea, one could hardly doubt that it was developed there by some purely natural forces. The discovery was a startling one, for it seemed that the actual starting point of life had been reached. Huxley named his substance Bathybias, and this name became in a short time familiar to every one who was thinking of the problems of life. But the discovery was suspected from the first, because it was too closely in accord with speculation, and it was soon disproved. Its discoverer soon after courageously announced to the world that he had been entirely mistaken, and that the Bathybias, so far from being undifferentiated protoplasm, was not an organic product at all, but simply a mineral deposit in the sea water made by purely artificial means. Bathybias stands therefore as an instance of a too precipitate advance in speculation, which led even such a brilliant man as Prof. Huxley into an unfortunate error of observation; for, beyond question, he would never have made such a mistake had he not been dominated by his speculative theories as to the nature of protoplasm.

But although Bathybias proved delusive, this did not materially affect the advance and development of the doctrine of protoplasm. Simple forms of protoplasm were found, although none quite so simple as the hypothetical Bathybias. The universal presence of protoplasm in the living parts of all animals and plants and its manifest activities completely demonstrated that it was the only living substance, and as the result of a few years of experiment and thought the biologist's conception of life crystallized into something like this: Living organisms are made of cells, but these cells are simply minute independent bits of protoplasm. They may contain a nucleus or they may not, but the essence of the cell is the protoplasm, this alone having the fundamental activities of life. These bits of living matter aggregate themselves together into groups to form colonies. Such colonies are animals or plants. The cells divide the work of the colony among themselves, each cell adopting a form best adapted for the special work it has to do. The animal or plant is thus simply an aggregate of cells, and its activities are the sum of the activities of its separate cells; just as the activities of a city are the sum of the activities of its individual inhabitants. The bit of protoplasm was the unit, and this was a chemical compound or a simple mixture of compounds to whose combined physical properties we have given the name vitality.

The Decline of the Reign of Protoplasm.—Hardly had this extreme chemical theory of life been clearly conceived before accumulating facts began to show that it is untenable and that it must at least be vastly modified before it can be received. The foundation of the chemical theory of life was the conception that protoplasm is a definite though complex chemical compound. But after a few years' study it appeared that such a conception of protoplasm was incorrect. It had long been suspected that protoplasm was more complex than was at first thought. It was not even at the outset found to be perfectly homogeneous, but was seen to contain minute granules, together with bodies of larger size. Although these bodies were seen they were regarded as accidental or secondary, and were not thought of as forming any serious objection to the conception of protoplasm as a definite chemical compound. But modern opticians improved their microscopes, and microscopists greatly improved their methods. With the new microscopes and new methods there began to appear, about twenty years ago, new revelations in regard to this protoplasm. Its lack of homogeneity became more evident, until there has finally been disclosed to us the significant fact that protoplasm is to be regarded as a substance not only of chemical but also of high mechanical complexity. The idea of this material as a simple homogeneous compound or as a mixture of such compounds is absolutely fallacious. Protoplasm is to-day known to be made up of parts harmoniously adapted to each other in such a way as to form an extraordinarily intricate machine; and the microscopist of to-day recognizes clearly that the activities of this material must be regarded as the result of the machinery which makes up protoplasm rather than as the simple result of its chemical composition. Protoplasm is a machine and not a chemical compound.

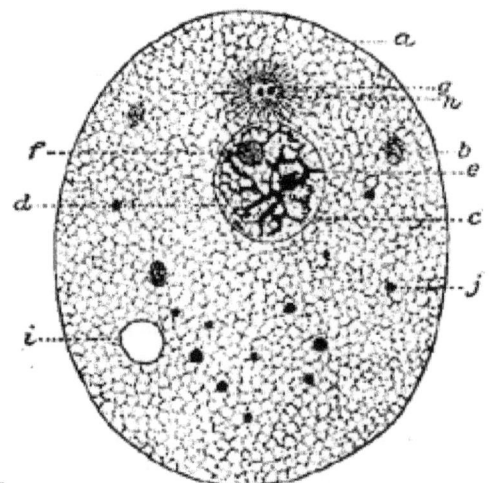

FIG. 23.—A cell as it appears to the modern microscope. a, protoplasmic reticulum; b, liquid in its meshes; c, nuclear membrane; d, nuclear reticulum; e, chromatin reticulum; f, nucleolus; g, centrosome; h, centrosphere; i, vacuole; j, inert bodies.

Structure of Protoplasm.—The structure of protoplasm is not yet thoroughly understood by scientists, but a few general facts are known beyond question. It is thought, in the first place, that it consists of two quite different substances. There is a somewhat solid material permeating it, usually, regarded as having a reticulate structure. It is variously described, sometimes as a reticulate network, sometimes as a mass of threads or fibres, and sometimes as a mass of foam (Fig. 23, a). It is extremely delicate and only visible under special conditions and with the best of microscopes. Only under peculiar conditions can it be seen in protoplasm while alive. There is no question, however, that all protoplasm is permeated when alive by a minute delicate mass of material, which may take the form of threads or fibres or may assume other forms. Within the meshes of this thread or reticulum there is found a liquid, perfectly clear and transparent, to whose presence the liquid character of the protoplasm is due (Fig. 23, b). In this liquid no structure can be determined, and, so far as we know, it is homogeneous. Still further study discloses other complexities. It appears that the fibrous material is always marked by the presence of excessively minute bodies, which have been called by various names, but which we will speak of as microsomes. Sometimes, indeed, the fibres themselves appear almost like strings of beads, so that they have been described as made up of rows of minute elements. It is immaterial for our purpose, however, whether the fibres are to be regarded as made up of microsomes or not. This much is sure, that these microsomes —granules of excessive minuteness—occur in protoplasm and are closely connected with the fibres (Fig. 23, a).

The Nucleus.—(a) Presence of a Nucleus.—If protoplasm has thus become a new substance in our minds as the result of the discoveries of the last twenty years, far more marvelous have been the discoveries made in connection with that body which has been called the nucleus. Even by the early microscopists the nucleus was recognized, and during the first few years of the cell doctrine it was frequently looked upon as the most active part of the cell and as especially connected with its reproduction. The doctrine of protoplasm, however, so captivated the minds of biologists that for quite a number of years the nucleus was ignored, at least in all discussions connected with the nature of life. It was a body in the cell whose presence was unexplained and which did not fall into accord with the general view of protoplasm as the physical basis of life. For a while, therefore, biologists gave little attention to it, and were accustomed to speak of it simply as a bit of protoplasm a little more dense than the rest. The cell was a bit of protoplasm with a small piece of more dense protoplasm in its centre appearing a little different from the rest and perhaps the most active part of the cell.

As a result of this excessive belief in the efficiency of protoplasm the question of the presence of a nucleus in the cell was for a while looked upon as one of comparatively little importance. Many cells were found to have nucleii while others did not show their presence, and microscopists therefore believed that the presence of a nucleus was not necessary to constitute a cell. A German naturalist recognized among lower animals one group whose distinctive characteristic was that they were made of cells without nucleii, giving the name Monera to the group. As the method of studying cells improved microscopists learned better methods of discerning the presence of the nucleus, and as it was done little by little they began to find the presence of nucleii in cells in which they had hitherto not been seen.

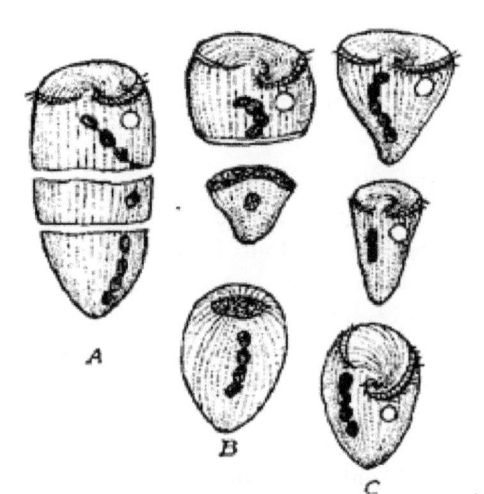

FIG. 24.—A cell cut into three pieces, each containing a bit of the nucleus. Each continues its life indefinitely, soon acquiring the form of the original as at C.

As microscopists now studied one after another of these animals and plants whose cells had been said to contain no nucleus, they began to find nucleii in them, until the conclusion was finally reached that a nucleus is a fundamental part of all active cells. Old cells which have lost their activity may not show nucleii, but, so far as we know, all active cells possess these structures, and apparently no cell can carry on its activity without them. Some cells have several nucleii, and others have the nuclear matter scattered through the whole cell instead of being aggregated into a mass; but nuclear matter the cell must have to carry on its life.

Later the experiment was made of depriving cells of their nucleii, and it still further emphasized the importance of the nucleus. Among unicellular animals are some which are large enough for direct manipulation, and it is found that if these cells are cut into pieces the different pieces will behave very differently in accordance with whether or not they have within them a piece of the nucleus. All the pieces are capable of carrying on their life activities for a while.

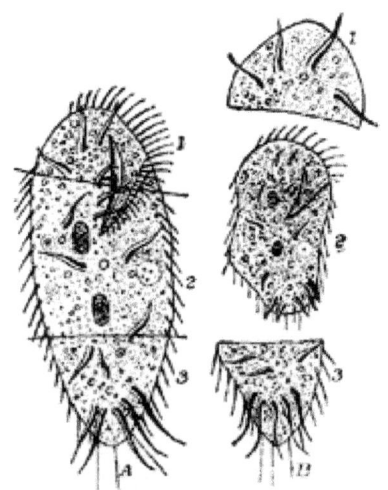

FIG. 25.—A cell cut into three pieces, only one of which, No. 2, contains any nucleus. This fragment soon acquires the original form and continues its life indefinitely, as shown at B. The other two pieces though living for a time, die without reproducing.

The pieces of the cell which contain the nucleus of the original cell, or even a part of it, are capable of carrying on all its life activities perfectly well. In Fig. 24 is shown such a cell cut into three pieces, each of which contains a piece of the nucleus. Each carries on its life activities, feeds, grows and multiplies perfectly well, the life processes seeming to continue as if nothing had happened. Quite different is it with fragments which contain none of the nucleus (Fig. 25). These fragments (1 and 3), even though they may be comparatively large masses of protoplasm, are incapable of carrying on the functions of their life continuously. For a while they continue to move around and apparently act like the other fragments, but after a little their life ceases. They are incapable of assimilating food and incapable of reproduction, and hence their life cannot continue very long. Facts like these demonstrate conclusively the vital importance of the nucleus in cell activity, and show us that the cell, with its power of continued life, must be regarded as a combination of protoplasm with its nucleus, and cannot exist without it. It is not protoplasm, but cell substance, plus cell nucleus, which forms the simplest basis of life.

As more careful study of protoplasm was made it soon became evident that there is a very decided difference between the nucleus and the protoplasm. The old statement that the nucleus is simply a bit of dense protoplasm is not true. In its chemical and physical composition as well as in its activities the nucleus shows itself to be entirely different from the protoplasm. It contains certain definite bodies not found in the cell substance, and it goes through a series of activities which are entirely unrepresented in the surrounding protoplasm. It is something entirely distinct, and its relations to the life of the cell are unique and marvelous. These various facts led to a period in the discussion of biological topics which may not inappropriately be called the Reign of the Nucleus. Let us, therefore, see what this structure is which has demanded so much attention in the last twenty years.

(b) Structure of the Nucleus.—At first the nucleus appears to be very much like the cell substance. Like the latter, it is made of fibres, which form a reticulum (Fig. 23), and these fibres, like those of protoplasm, have microsomes in intimate relation with them and hold a clear liquid in their meshes. The meshes of the network are usually rather closer than in the outer cell substance, but their general character appears to be the same. But a more close study of the nucleus discloses vast differences. In the first place, the nucleus is usually separated from the cell substance by a membrane (Fig. 23, c). This membrane is almost always present, but it may disappear, and usually does disappear, when the nucleus begins to divide. Within the nucleus we find commonly one or two smaller bodies, the nucleoli (Fig. 23, f). They appear to be distinct vital parts of the nucleus, and thus different from certain other solid bodies which are simply excreted material, and hence lifeless. Further, we find that the reticulum within the nucleus is made up of two very different parts. One portion is apparently identical with the reticulum of the cell substance (Fig. 23, d). This forms an extremely delicate network, whose fibres have chemical relations similar to those of the cell substance. Indeed, sometimes, the fibres of the nucleus may be seen to pass directly into those of the network of the cell substance, and hence they are in all probability identical. This material is called linin, by which name we shall hereafter refer to it. There is, however, in the nucleus another material which forms either threads, or a network, or a mass of granules, which is very different from the linin, and has entirely different properties. This network has the power of absorbing certain kinds of stains very actively, and is consequently deeply stained when treated as the microscopist commonly prepares his specimens. For this reason it has been named chromatin (Fig, 23, e), although in more recent times other names have been given to it. Of all parts of the cell this chromatin is the most remarkable. It appears in great variety in different cells, but it always has remarkable physiological properties, as will be noticed presently. All things considered, this chromatin is probably the most remarkable body connected with organic life.

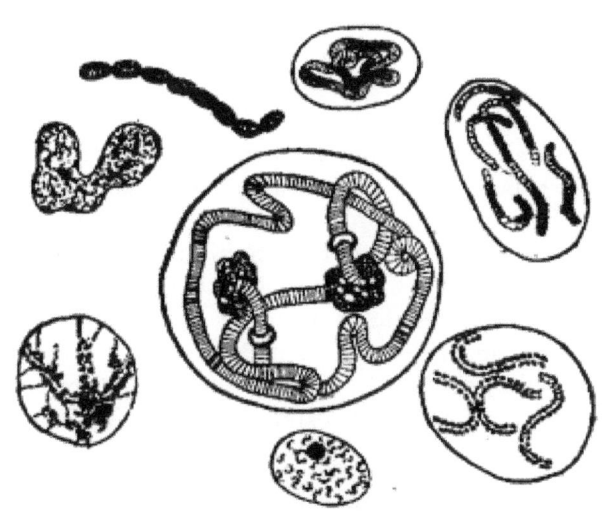

FIG. 26.—Different forms of nucleii.

The nucleii of different animals and plants all show essentially the characteristics just described. They all contain a liquid, a linin network, and a chromatin thread or network, but they differ most remarkably in details, so that the variety among the nucleii is almost endless (Fig. 26). They differ first in their size relative to the size of the cell; sometimes—especially in young cells—the nucleus being very large, while in other cases the nucleus is very small and the protoplasmic contents of the cell very large; finally, in cells which have lost their activity the nucleus may almost or entirely disappear. They differ, secondly, in shape. The typical form appears to be spherical or nearly so; but from this typical form they may vary, becoming irregular or elongated. They are sometimes drawn out into long masses looking like a string of beads (Fig. 24), or, again, resembling minute coiled worms (Fig. 21), while in still other cells they may be branching like the twigs of a tree. The form and shape of the chromatin thread differs widely. Sometimes this appears to be mere reticulum (Fig. 23); at others, a short thread which is somewhat twisted or coiled (Fig. 26); while in other cells the chromatin thread is an extremely long, very much twisted convolute thread so complexly woven into a tangle as to give the appearance of a minute network. The nucleii differ also in the number of nucleoli they contain as well as in other less important particulars. Fig. 26 will give a little notion of the variety to be found among different nucleii; but although they thus do vary most remarkably in shape in the essential parts of their structure they are alike.

Centrosome.—Before noticing the activities of the nucleus it will be necessary to mention a third part of the cell. Within the last few years there has been found to be present in most cells an organ which has been called the centrosome. This body is shown at Fig. 23, g. It is found in the cell substance just outside the nucleus, and commonly appears as an extremely minute rounded dot, so minute that no internal structure has been discerned. It may be no larger than the minute granules or microsomes in the cell, and until recently it entirely escaped the notice of microscopists. It has now, however, been clearly demonstrated as an active part of the cell and entirely distinct from the ordinary microsomes. It stains differently, and, as we shall soon see, it appears to be in most intimate connection with the center of cell life. In the activities which characterize cell life this centrosome appears to lead the way. From it radiate the forces which control cell activity, and hence this centrosome is sometimes called the dynamic center of the cell. This leads us to the study of cell activity, which discloses to us some of the most extraordinary phenomena which have come to the knowledge of science.

Function of the Nucleus.—To understand why it is that the nucleus has taken such a prominent position in modern biological discussion it will be only necessary to notice some of the activities of the cell. Of the four fundamental vital properties of cell life the one which has been most studied and in regard to which most is known is reproduction. This knowledge appears chiefly under two heads, viz., cell division and the fertilization of the egg. Every animal and plant begins its life as a simple cell, and the growth of the cell into the adult is simply the division of the original cell into parts accompanied by a differentiation of the parts. The fundamental phenomena of growth and reproduction is thus cell division, and if we can comprehend this process in these simple cells we shall certainly have taken a great step toward the explanation of the mechanics of life. During the last ten years this cell division has been most thoroughly studied, and we have a pretty good knowledge of it so far as its microscopical features are concerned. The following description will outline the general facts of such cell division, and will apply with considerable accuracy to all cases of cell division, although the details may differ not a little.

Cell Division or Karyokinesis.—We will begin with a cell in what is called the resting stage, shown at Fig. 23. Such a cell has a nucleus, with its chromatin, its membrane, and linin, as already described. Outside the nucleus is the centrosome, or, more commonly, two of them lying close together. If there is only one it soon divides into two, and if it has already two, this is because a single centrosome which the cell originally possessed has already divided into two, as we shall presently see. This cell, in short, is precisely like the typical cell which we have described, except in the possession of two centrosomes.

FIG. 27.   FIG. 28.

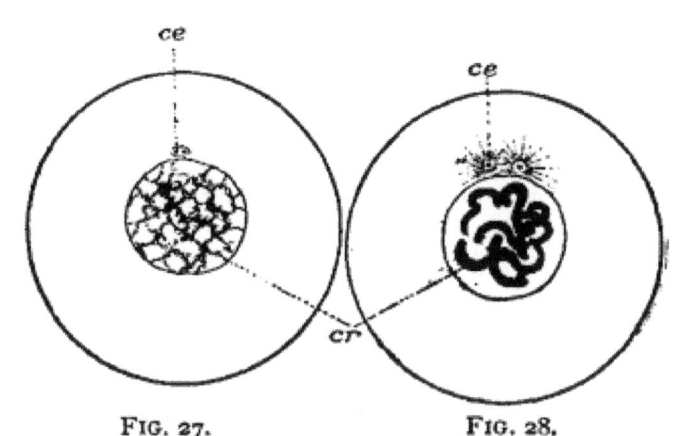

FIG. 27 shows the resting stage with the chromatin, cr, in the form of a network within the nuclear membrane and the centrosome, ce, already divided into two.

FIG. 28.—The chromatin is broken into threads or chromosomes, cr. The centrosomes show radiating fibres.

The first indication of the cell division is shown by the chromatin fibres. During the resting stage this chromatin material may have the form of a thread, or may form a network of fibres (see Fig. 27). But whatever be its form during the resting stage, it assumes the form of a thread as the cell prepares for division. Almost at once this thread breaks into a number of pieces known as chromosomes (Fig. 28). It is an extremely important fact that the number of these chromosomes in the ordinary cells of any animal or plant is always the same. In other words, in all the cells of the body of animal or plant the chromatin material in the nucleus breaks into the same number of short threads at the time that the cell is preparing to divide. The number is the same for all animals of the same species, and is never departed from. For example, the number in the ox is always sixteen, while the number in the lily is always twenty-four. During this process of the formation of the chromosomes the nucleoli disappear, sometimes being absorbed apparently in the chromosomes, and sometimes being ejected into the cell body, where they disappear. Whether they have anything to do with further changes is not yet known.

The next step in the process of division appears in the region of the centrosomes. Each of the two centrosomes appears to send out from itself delicate radiating fibres into the surrounding cell substance (Fig. 28). Whether these actually arise from the centrosome or are simply a rearrangement of the fibres in the cell substance is not clear, but at all events the centrosome becomes surrounded by a mass of radiating fibres which give it a starlike appearance, or, more commonly, the appearance of a double star, since there are two centrosomes close together (Fig. 28). These radiating fibres, whether arising from the centrosomes or not, certainly all centre in these bodies, a fact which indicates that the centrosomes contain the forces which regulate their appearance. Between the two stars or asters a set of fibres can be seen running from one to the other (Fig. 29). These two asters and the centrosomes within them have been spoken of as the dynamic centre of the cell since they appear to control the forces which lead to cell division. In all the changes which follow these asters lead the way. The two asters, with their centrosomes, now move away from each other, always connected by the spindle fibres, and finally come to lie on opposite sides of the nucleus (Figs. 29, 30). When they reach this position they are still surrounded by the radiating fibres, and connected by the spindle fibres. Meantime the membrane around the nucleus has disappeared, and thus the spindle fibres readily penetrate into the nuclear substance (Fig. 30).

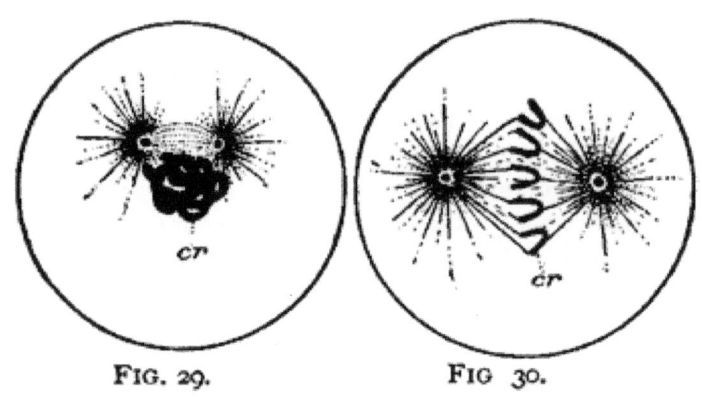

FIG. 29.   FIG 30.

FIG. 29.—The centrosomes are separating but are connected by fibres.
FIG. 30.—The centrosomes are separate and the equatorial plate of chromosomes, cr, is between them.

During this time the chromosomes have been changing their position. Whether this change in position is due to forces within themselves, or whether they are moved around passively by forces residing in the cell substances, or whether, which is the most probable, they are pulled or pushed around by the spindle fibres which are forcing their way into the nucleus, is not positively known; nor is it, for our purposes, of special importance. At all events, the result is that when the asters have assumed their position at opposite poles of the nucleus the chromosomes are arranged in a plane passing through the middle of the nucleus at equal distances from each aster. It seems certain that they are pulled or pushed into this position by forces radiating from the centrosomes. Fig. 30 shows this central arrangement of the chromosomes, forming what is called the equatorial plate.

The next step is the most significant of all. It consists in the splitting of each chromosome into two equal halves. The threads do not divide in their middle but split lengthwise, so that there are formed two halves identical in every respect. In this way are produced twice the original number of chromosomes, but all in pairs. The period at which this splitting of the chromosomes occurs is not the same in all cells. It may occur, as described, at about the time the asters have reached the opposite poles of the nucleus, and an equatorial plate is formed. It is not infrequent, however, for it to occur at a period considerably earlier, so that the chromosomes are already divided when they are brought into the equatorial plate.

At some period or other in the cell division this splitting of the chromosomes takes place. The significance of the splitting is especially noteworthy. We shall soon find reason for believing that the chromosomes contain all the hereditary traits which the cell hands down from generation to generation, and indeed that the chromosomes of the egg contain all the traits which the parent hands down to the child. Now, if this chromatin thread consists of a series of units, each representing certain hereditary characters, then it is plain that the division of the thread by splitting will give rise to a double series of threads, each of which has identical characters. Should the division occur across the thread the two halves would be unlike, but taking place as it does by a longitudinal splitting each unit in the thread simply divides in half, and thus the resulting half threads each contain the same number of similar units as the other and the same as possessed by the original undivided chromosome. This sort of splitting thus doubles the number of chromosomes, but produces no differentiation of material.

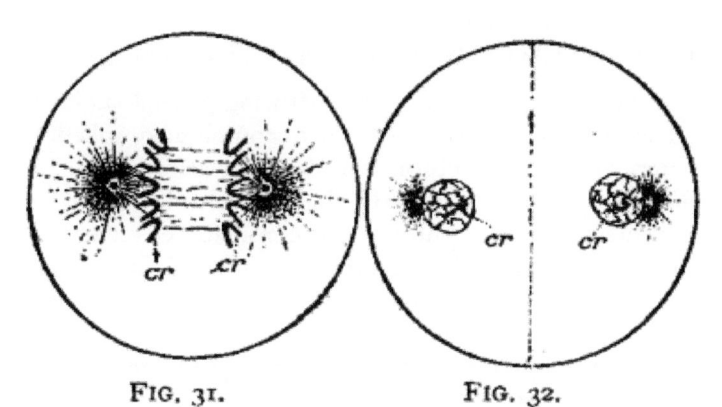

FIG. 31.   FIG. 32.
FIG. 31.—Stage showing the two halves of the chromosomes separated from each other.

FIG. 32.—Final stage with two nucleii in which the chromosomes have again assumed the form of a network. The centrosomes have divided preparatory to the next division, and the cell is beginning to divide.

The next step in the cell division consists in the separation of the two halves of the chromosomes. Each half of each chromosome separates from its fellow, and moves to the opposite end of the nucleus toward the two centrosomes (Fig. 31). Whether they are pulled apart or pushed apart by the spindle fibres is not certain, although it is apparently sure that these fibres from the centrosomes are engaged in the matter. Certain it is that some force exerted from the two centrosomes acts upon the chromosomes, and forces the two halves of each one to opposite ends of the nucleus, where they now collect and form two new nucleii, with evidently exactly the same number of chromosomes as the original, and with characters identical to each other and to the original (Fig. 32).

The rest of the cell division now follows rapidly. A partition grows in through the cell body dividing it into two parts (Fig. 32), the division passing through the middle of the spindle. In this division, in some cases at least, the spindle fibres bear a part—a fact which again points to the importance of the centrosomes and the forces which radiate from them. Now the chromosomes in each daughter nucleus unite to form a single thread, or may diffuse through the nucleus to form a network, as in Fig. 32. They now become surrounded by a membrane, so that the new nucleus appears exactly like the original one. The spindle fibres disappear, and the astral fibres may either disappear or remain visible. The centrosome may apparently in some cases disappear, but more commonly remains beside the daughter nucleii, or it may move into the nucleus. Eventually it divides into two, the division commonly occurring at once (Fig. 32), but sometimes not until the next cell division is about to begin. Thus the final result shows two cells each with a nucleus and two centrosomes, and this is exactly the same sort of structure with which the process began. (See Frontispiece.)

Viewed as a whole, we may make the following general summary of this process. The essential object of this complicated phenomena of karyokinesis is to divide the chromatin into equivalent halves, so that the cells resulting from the cell division shall contain an exactly equivalent chromatin content. For this purpose the chromatic elements collect into threads and split lengthwise. The centrosome, with its fibres, brings about the separation of these two halves. Plainly, we must conclude that the chromatin material is something of extraordinary importance to the cell, and the centrosome is a bit of machinery for controlling its division and thus regulating cell division.

Fertilization of the Egg.—This description of cell division will certainly give some idea of the complexity of cell life, but a more marvelous series of changes still takes place during the time when the egg is preparing for development. Inasmuch as this process still further illustrates the nature of the cell, and has further a most intimate bearing upon the fundamental problem of heredity, it will be necessary for us to consider it here briefly.

The sexual reproduction of the many-celled animals is always essentially alike. A single one of the body cells is set apart to start the next generation, and this cell, after separating from the body of the animal or plant which produced it, begins to divide, as already shown in Fig. 8, and the many cells which arise from it eventually form the new individual This reproductive cell is the egg. But before its division can begin there occurs in all cases of sexual reproduction a process called fertilization, the essential feature of which is the union of this cell with another commonly from a different individual. While the phenomenon is subject to considerable difference in details, it is essentially as follows:

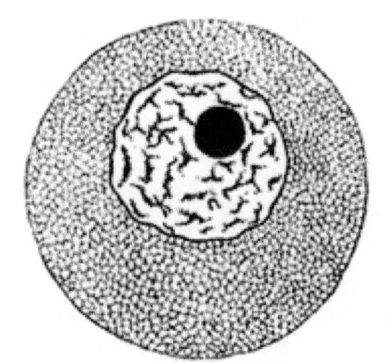

FIG. 33—An egg showing the cell substance and the nucleus, the latter containing chromosomes in large number and a nucleolus.

The female reproductive cell is called the egg, and it is this cell which divides to form the next generation. Such a cell is shown in Fig. 33. Like other cells it has a cell wall, a cell substance with its linin and fluid portions, a nucleus surrounded by a membrane and containing a reticulum, a nucleolus and chromatic material, and lastly, a centrosome. Now such an egg is a complete cell, but it is not able to begin the process of division which shall give rise to a new individual until it has united with another cell of quite a different sort and commonly derived from a different individual called the male. Why the egg cell is unable to develop without such union with male cell does not concern us here, but its purpose will be evident as the description proceeds. The egg cell as it comes from the ovary of the female individual is, however, not yet ready for union with the male cell, but must first go through a series of somewhat remarkable changes constituting what is called maturation of the egg. This phenomenon has such an intimate relation to all problems connected with the cell, that it must be described somewhat in detail. There are considerable differences in the details of the process as it occurs in various animals, but they all agree in the fundamental points. The following is a general description of the process derived from the study of a large variety of animals and plants.

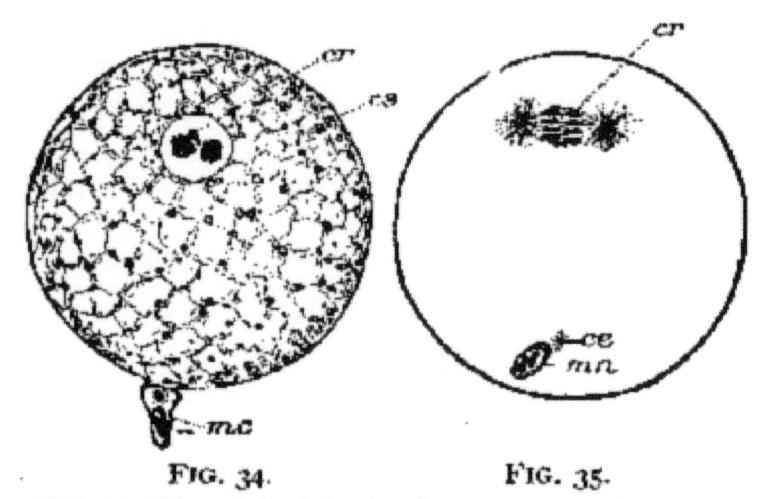

FIG. 34.   FIG. 35.

FIG. 34. This and the following figures represent the process of fertilization of an egg. In all figures cr is the chromosomes; cs represents the cell substance (omitted in the following figures); mc is the male reproductive cell lying in contact with the egg; mn is the male nucleus after entering the egg.

FIG. 35.—The egg centrosome has divided, and the male cell with its centrosome has entered the egg.

In the cells of the body of the animal to which this description applies there are four chromosomes This is true of all the cells of the animal except the sexual cells. The eggs arise from the other cells of the body, but during their growth the chromatin splits in such a way that the egg contains double the number of chromosomes, i.e., eight (Fig. 34). If this egg should now unite with the other reproductive cell from the male, the resulting fertilized egg would plainly contain a number of chromosomes larger than that normal for this species of animal. As a result the next generation would have a larger number of chromosomes in each cell than the last generation, since the division of the egg in development is like that already described and always results in producing new cells with the same number of chromosomes as the starting cell. Hence, if the number of chromosomes in the next generation is to be kept equal to that in the last generation, this egg cell must get rid of a part of its chromatin material.

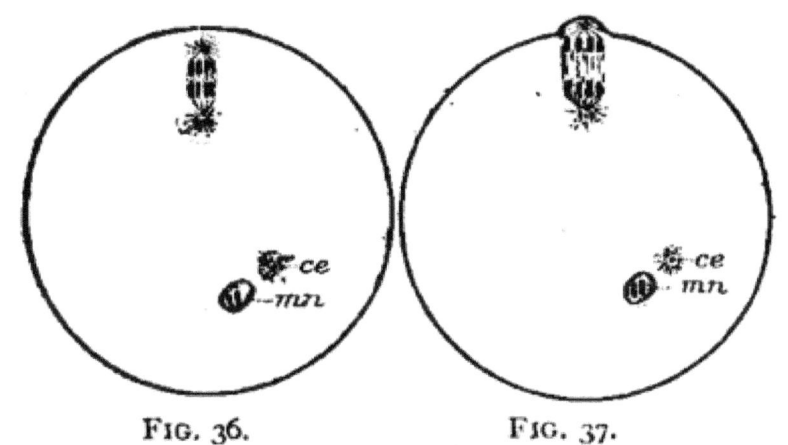

FIG. 36.   FIG. 37.

FIG. 36—The egg centrosomes have changed their position. The male cell with its centrosome remains inactive until the stage represented in FIG. 42.

FIG. 37—Beginning of the first division for removing superfluous chromosomes.

This is done by a process shown in Fig. 35. The centrosome divides as in ordinary cell division (Fig. 35), and after rotating on its axis it approaches the surface of the egg (Figs. 36 and 37). The egg now divides (Fig. 38), but the division is of a peculiar kind. Although the chromosomes divide equally the egg itself divides into two very unequal parts, one part still appearing as the egg and the other as a minute protuberance called the polar cell (pc' in Fig. 38). The chromosomes do not split as they do in the cell division already described, but each of these two cells, the egg and the polar body, receives four chromosomes (Fig. 38). The result is that the egg has now the normal number of chromosomes for the ordinary cells of the animal in question. But this is still too many, for the egg is soon to unite with the male cell; and this male cell, as we shall see, is to bring in its own quota of chromosomes. Hence the egg must get rid of still more of its chromatin material. Consequently, the first division is followed by a second (Fig. 39), in which there is again produced a large and a small cell. This division, like the first, occurs without any splitting of the chromosomes, one half of the remaining chromosomes being ejected in this new cell, the second polar cell (pc") leaving the larger cell, the egg, with just one half the number of chromosomes normal for the cells of the animal in question. Meantime the first pole cell has also divided, so that we have now, as shown in Fig. 40, four cells, three small and one large, but each containing one half the normal number of chromosomes. In the example figured, four is the normal number for the cells of the animal. The egg at the beginning of the process contained eight, but has now been reduced to two. In the further history of the egg the smaller cells, called polar cells, take no part, since they soon disappear and have nothing to do with the animal which is to result from the further division of the egg. This process of the formation of the polar cells is thus simply a device for getting rid of some of the chromatin material in the egg cell, so that it may unite with a second cell without doubling the normal number of chromosomes.

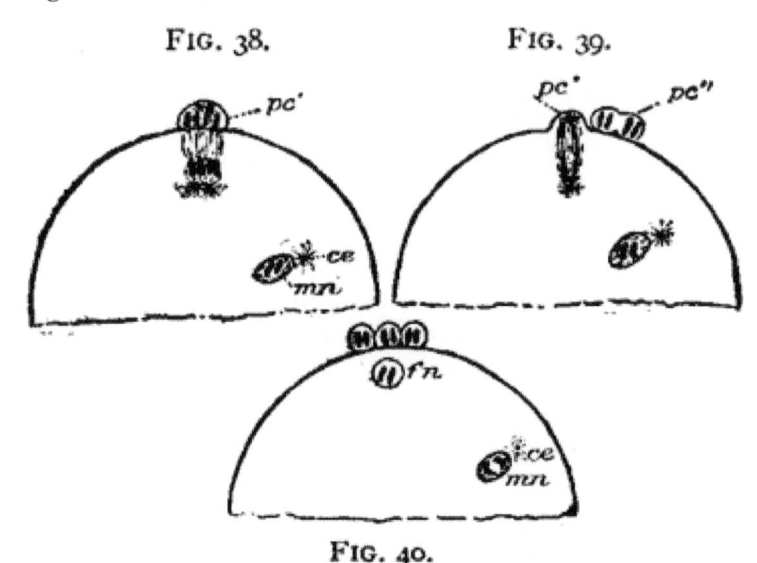

FIG. 38.

FIG. 39.

FIG. 40.

FIG.38—First division complete and first polar cell formed, pc'.
FIG.39.—Formation of the second polar cell, pc".
FIG.40.—Completion of the process of extrusion of the chromatic material; fn shows the two chromosomes retained in the egg forming the female pronucleus. The centrosome has disappeared.

Previously to this process the other sexual cell, the spermatozoon, or male reproductive cell, has been undergoing a somewhat similar process. This is also a true cell (Fig. 34, mc), although it is of a decidedly smaller size than the egg and of a very different shape. It contains cell substance, a nucleus with chromosomes, and a centrosome, the number of chromosomes, as shown later, being however only half that normal for the ordinary cells of the animals. The study of the development of the spermatozoon shows that it has come from cells which contained the normal number of four, but that this number has been reduced to one half by a process which is equivalent to that which we have just noticed in the egg. Thus it comes about that each of the sexual elements, the egg and the spermatozoon, now contains one half the normal number of chromosomes.

Now by some mechanical means these two reproductive cells are brought in contact with each other, shown in Fig. 34, and as soon as they are brought into each other's vicinity the male cell buries its head in the body of the egg. The tail by which it has been moving is cast off, and the head containing the chromosomes and the centrosome enters the egg, forming what is called the male pronucleus (Fig. 35-38, mn). This entrance of the male cell occurs either before the formation of the polar cells of the egg or afterward. If, however, it takes place before, the male pronucleus simply remains dormant in the egg while the polar cells are being protruded, and not until after that process is concluded does it begin again to show signs of activity which result in the cell union.

The further steps in this process appear to be controlled by the centrosome, although it is not quite certain whence this centrosome is derived. Originally, as we have seen, the egg contained a centrosome, and the male cell has also brought a second into the egg (Fig. 35, ce). In some cases, and this is true for the worm we are describing, it is certain that the egg centrosome disappears while that of the spermatozoon is retained alone to direct the further activities (Fig. 41). Possibly this may be the case in all eggs, but it is not sure. It is a matter of some little interest to have this settled, for if it should prove true, then it would evidently follow that the machinery for cell division, in the case of sexual reproduction, is derived from the father, although the bulk of the cell comes from the mother, while the chromosomes come from both parents.

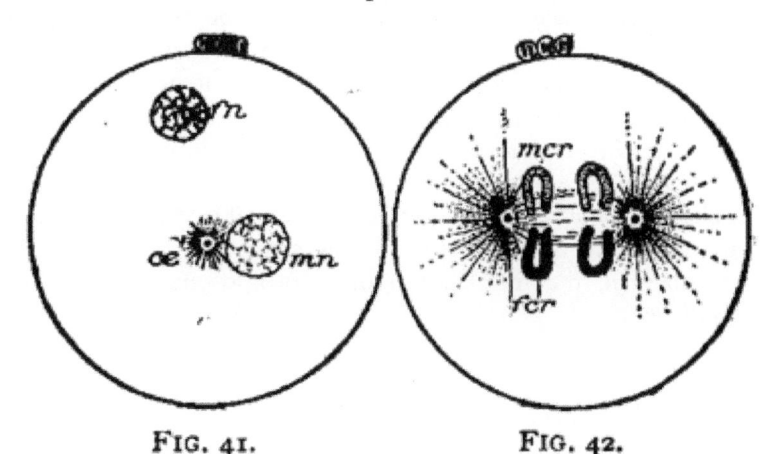

FIG. 41. FIG. 42.

FIG. 41.—The chromosomes in the male and female pronucleii have resolved into a network. The male centrosome begins to show signs of activity.

FIG. 42.—The centrosome has divided, and the two pronucleii have been brought together. The network in each nucleus has again resolved itself into two chromosomes which are now brought together near the centre of the egg but do not fuse; mcr, represents the chromosomes from the male nucleus; fcr, the chromosomes from the female nucleus.

In the cases where the process has been most carefully studied, the further changes are as follows: The head of the spermatozoon, after entrance into the egg, lies dormant until the egg has thrown off its polar cells, and thus gotten rid of part of its chromosomes. Close to it lies its centrosomes (Fig. 35, ce), and there is thus formed what is known as the male pronucleus (Fig. 35-40, mn). The remains of the egg nucleus, after having discharged the polar cells, form the female nucleus (Fig. 40, fn). The chromatin material, in both the male and female pronucleus, soon breaks up into a network in which it is no longer possible to see that each contains two chromosomes (Fig. 41). Now the centrosome, which is beside the male pronucleus, shows signs of activity. It becomes surrounded by prominent rays to form an aster (Fig. 41, ce), and then it begins to move toward the female pronucleus, apparently dragging the male pronucleus after it. In this way the centrosome approaches the female pronucleus, and thus finally the two nucleii are brought into close proximity. Meantime the chromatin material in each has once more broken up into short threads or chromosomes, and once more we find that each of the nucleii contains two of these bodies (Fig. 42). In the subsequent figures the chromosomes of the male nucleus are lightly shaded, while those of the female are black in order to distinguish them. As these two nucleii finally come together their membranes disappear, and the chromatic material comes to lie freely in the egg, the male and female chromosomes, side by side, but distinct forming the segmentation nucleus. The egg plainly now contains once more the number of chromosomes normal for the cells of the animal, but half of them have been derived from each parent. It is very suggestive to find further that the chromosomes in this fertilized egg do not fuse with each other, but remain quite distinct, so that it can be seen that the new nucleus contains chromosomes derived from each parent (Fig. 42). Nor does there appear to be, in the future history of this egg, any actual fusion of the chromatic material, the male and female chromosomes perhaps always remaining distinct.

FIG. 43.   FIG. 44.

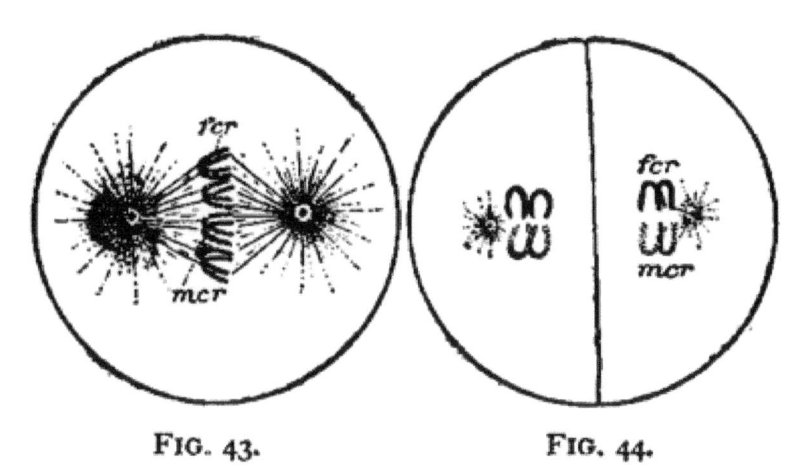

FIG. 43.—An equatorial plate is formed and each chromosome has split into two halves by longitudinal division.

FIG. 44.—The halves of the chromosomes have separated to form two nucleii, each with male and female chromosomes. The egg has divided into two cells.

While this mixture of chromosomes has been taking place the centrosome has divided into two parts, each of which becomes surrounded by an aster and travels to opposite ends of the nucleus (Fig. 42). There now follows a division of the nucleus exactly similar to that which occurs in the normal cell division already described in Figs. 28-34. Each of the chromosomes splits lengthwise (Fig. 43), and one half of each then travels toward each centrosome to form a new nucleus (Fig. 44). Since each of the four chromosomes thus splits, it follows that each of the two daughter nucleii will, of course, contain four chromosomes; two of which have been derived from the male and two from the female parent. From now the divisions of the egg follow rapidly by the normal process of cell division until from this one egg cell there are eventually derived hundreds of thousands of cells which are gradually moulded into the adult. All of these cells will, of course, contain four chromosomes; and, what is more important, half of the chromosomes will have been derived directly from the male and half from the female parent. Even into adult life, therefore, the cells of the animal probably contain chromatin derived by direct descent from each of its parents.

The Significance of Fertilization.—From this process of fertilization a number of conclusions, highly important for our purpose, can be drawn. In the first place, it is evident that the chromosomes form the part of the cell which contain the hereditary traits handed down from parent to child. This follows from the fact that the chromosomes are the only part of the cell which, in the fertilized egg, is derived from both parents. Now the offspring can certainly inherit from each parent, and hence the hereditary traits must be associated with some part of the cell which is derived from both. But the egg substance is derived from the mother alone; the centrosome, at least in some cases and perhaps in all, is derived only from the father, while the chromosomes are derived from both parents. Hence it follows that the hereditary traits must be particularly associated with the chromosomes.

With this understanding we can, at least, in part understand the purpose of fertilization. As we shall see later, it is very necessary in the building of the living machine for each individual to inherit characters from more than one individual. This is necessary to produce the numerous variations which contribute to the construction of the machine. For this purpose there has been developed the process of sexual union of reproductive cells, which introduces into the offspring chromatic material from two parents. But if the two reproductive cells should unite at once the number of chromosomes would be doubled in each generation, and hence be constantly increasing. To prevent this the polar cells are cast out, which reduces the amount of chromatic material. The union of the two pronucleii is plainly to produce a nucleus which shall contain chromosomes, and hence hereditary traits from each parent and the subsequent splitting of these chromosomes and the separation of the two halves into daughter nucleii insures that all the nucleii, and hence all cells of the adult, shall possess hereditary traits derived from both parents. Thus it comes that, even in the adult, every body cell is made up of chromosomes from each parent, and may hence inherit characters from each.

The cell of an animal thus consists of three somewhat distinct but active parts—the cell substance, the chromosomes, and the centrosome. Of these the cell substance appears to be handed down from the mother; the centrosome comes, at least in some cases, from the father, and the chromosomes from both parents. It is not yet certain, however, whether the centrosome is a constant part of the cell. In some cells it cannot yet be found, and there are some reasons for believing that it may be formed out of other parts of the cell. The nucleus is always a direct descendant from the nucleus of pre-existing cells, so that there is an absolute continuity of descent between the nucleii of the cells of an individual and those of its antecedents back for numberless generations. It is not certain that there is any such continuity of descent in the case of the centrosomes; for, while in the process of fertilization the centrosome is handed down from parent to child, there are some reasons for believing that it may disappear in subsequent cells, and later be redeveloped out of other parts. The only part of the cell in which complete continuity from parent to child is demonstrated, is the nucleus and particularly the chromosomes. All of these facts simply emphasize the importance of the chromosomes, and tell us that these bodies must be regarded as containing the most important features of the cell which constitute its individuality.

What is Protoplasm?—Enough has now been given of disclosures of the modern microscope to show that our old friend Protoplasm has assumed an entirely new guise, if indeed it has not disappeared altogether. These simplest life processes are so marvelous and involve the action of such an intricate mass of machinery that we can no longer retain our earlier notion of protoplasm as the physical basis of life. There can be no life without the properties of assimilation, growth, and reproduction; and, so far as we know, these properties are found only in that combination of bodies which we call the cell, with its mixture of harmoniously acting parts. Life, at least the life of a cell, is then not the property of a chemical compound protoplasm, but is the result of the activities of a machine. Indeed, we are now at a loss to know how we can retain the term protoplasm. As originally used it meant the contents of the cell, and the significance in the term was in the conception of protoplasm as a somewhat homogeneous chemical compound uniform in all types of life. But we now see that this cell contains not a single substance, but a large number, including solids, jelly masses, and liquids, each of which has its own chemical composition. The number of chemical compounds existing in the material formerly called protoplasm no one knows, but we do know that they are many, and that the different substances are combined to form a physical structure. Which of these various bodies shall we continue to call protoplasm? Shall it be the linin, or the liquids, or the microsomes, or the chromatin threads, or the centrosomes? Which of these is the actual physical basis of life? From the description of cell life which we have given, it will be evident that no one of them is a material upon which our chemical biologists can longer found a chemical theory of life. That chemical theory of life, as we have seen, was founded upon the conception that the primitive life substance is a definite chemical compound. No such compound has been discovered, and these disclosures of the microscope of the last few years have been such as to lead us to abandon hope of ever discovering such a compound. It is apparently impossible to reduce life to any simpler basis than this combination of bodies which make up what was formerly called protoplasm. The term protoplasm is still in use with different meanings as used by different writers. Sometimes it is used to refer to the entire contents of the cell; sometimes to the cell substance only outside the nucleus. Plainly, it is not the protoplasm of earlier years.

With this conclusion one of our fundamental questions has been answered. We found in our first chapter that the general activities of animals and plants are easily reduced to the action of a machine, provided we had the fundamental vital powers residing in the parts of that machine. We then asked whether these fundamental properties were themselves those of a chemical compound or whether they were to be reduced to the action of still smaller machines. The first answer which biologists gave to this question was that assimilation, growth, and reproduction were the simple properties of a complex chemical compound. This answer was certainly incorrect. Life activities are exhibited by no chemical compound, but, so far as we know, only by the machine called the cell. Thus it is that we are again reduced to the problem of understanding the action of a machine. It may be well to pause here a moment to notice that this position very greatly increases the difficulties in the way of a solution of the life problem. If the physical basis of life had proved to be a chemical compound, the problem of its origin would have been a chemical one. Chemical forces exist in nature, and these forces are sufficient to explain the formation of any kind of chemical compound. The problem of the origin of the life substance would then have been simply to account for certain conditions which resulted in such chemical combination as would give rise to this physical basis of life. But now that the simplest substance manifesting the phenomena of life is found to be a machine, we can no longer find in chemical forces efficient causes for its formation. Chemical forces and chemical affinity can explain chemical compounds of any degree of complexity, but they cannot explain the formation of machines. Machines are the result of forces of an entirely different nature. Man can manufacture machines by taking chemical compounds and putting them together into such relations that their interaction will give certain results. Bits of iron and steel, for instance, are put together to form a locomotive, but the action of the locomotive depends, not upon the chemical forces which made the steel, but upon the relation of the bits of steel to each other in the machine. So far as we have had any experience, machines have been built under the guidance of intelligence which adapts the parts to each other. When therefore we find that the simplest life substance is a machine, we are forced to ask what forces exist in nature which can in a similar way build machines by the adjustment of parts to each other. But this topic belongs to the second part of our subject, and must be for the present postponed.

Reaction against the Cell Doctrine.—As the knowledge of cells which we have outlined was slowly acquired, the conception of the cell passed through various modifications. At first the cell wall was looked upon as the fundamental part, but this idea soon gave place to the belief that it was the protoplasm that was alive. Under the influence of this thought the cell doctrine developed into something like the following: The cell is simply a bit of protoplasm and is the unit of living matter. The bodies of all larger animals and plants are made up of great numbers of these units acting together, and the activities of the entire organism are simply the sum of the activities of its cells. The organism is thus simply the sum of the cells which compose it, and its activities the sum of the activities of the individual cells. As more facts were disclosed the idea changed slightly. The importance of the nucleus became more and more forcibly impressed upon microscopists, and this body came after a little into such prominence as to hide from view the more familiar protoplasm. The marvellous activities of the nucleus soon caused it to be regarded as the important part of the cell, while all the rest was secondary. The cell was now thought of as a bit of nuclear matter surrounded by secondary parts. The marvellous activities of the nucleus, and above all, the fact that the nucleus alone is handed down from one generation to the next in reproduction, all attested to its great importance and to the secondary importance of the rest of the cell.

This was the most extreme position of the cell doctrine. The cell was the unit of living action, and the higher animal or plant simply a colony of such units. An animal was simply an association together for mutual advantage of independent units, just as a city is an association of independent individuals. The organization of the animals was simply the result of the combination of many independent units. There was no activity of the organism as a whole, but only of its independent parts. Cell life was superior to organized life. Just as, in a city, the city government is a name given to the combined action of the individuals, so are the actions of organisms simply the combined action of their individual cells.

Against such an extreme position there has been in recent years a decided reaction, and to-day it is becoming more and more evident that such a position cannot be maintained. In the first place, it is becoming evident that the cell substance is not to be entirely obliterated by the importance of the nucleus. That the nucleus is a most important vital centre is clear enough, but it is equally clear that nucleus and cell substance must be together to constitute the life substance. The complicated structure of the cell substance, the decided activity shown by its fibres in the process of cell division, clearly enough indicate that it is a part of the cell which can not be neglected in the study of the life substance. Again the discovery of the centrosome as a distinct morphological element has still further added to the complexity of the life substance, and proved that neither nucleus nor cell substance can be regarded as the cell or as constituting life. It is true that we may not yet know the source of this centrosome. We do not know whether it is handed down from generation to generation like the nucleus, or whether it can be made anew out of the cell substance in the life of an ordinary cell. But this is not material to its recognition as an organ of importance in the cell activity. Thus the cell proves itself not to; be a bit of nuclear matter surrounded by secondary parts, but a community of several perhaps equally important interrelated members.

Another series of observations weakened the cell doctrine in an entirely different direction. It had been assumed that the body of the multicellular animal or plant was made of independent units. Microscopists of a few years ago began to suggest that the cells are in reality not separated from each other, but are all connected by protoplasmic fibres. In quite a number of different kinds of tissue it has been determined that fine threads of protoplasmic material lead from one cell to another in such a way that the cells are in vital connection. The claim has been made that there is thus a protoplasmic connection between all the cells of the body of the animal, and that thus the animal or plant, instead of consisting of a large number of separate independent cells, consists of one great mass of living matter which is aggregated into little centres, each commonly holding a nucleus. Such a conclusion is not yet demonstrated, nor is its significance very clear should it prove to be a fact; but it is plain that such suggestions quite decidedly modify the conception of the body as a community of independent cells.

There is yet another line of thought which is weakening this early conception of the cell doctrine. There is a growing conviction that the view of the organism, simply as the sum of the activities of the individual cells, is not a correct understanding of it. According to this extreme position, a living thing can have no organization until it appears as the result of cell multiplication. To take a concrete case, the egg of a starfish can not possess any organization corresponding to the starfish. The egg is a single cell, and the starfish a community of cells. The egg can, therefore, no more contain the organization of a starfish than a hunter in the backwoods can contain within himself the organization of a great metropolis. The descendants of individuals like the hunter may unite to form a city, and the descendants of the egg cell may, by combining, give rise to the starfish. But neither can the man contain within himself the organization of the city, nor the egg that of the starfish. It is, perhaps, true that such an extreme position of the cell doctrine has not been held by any one, but thoughts very closely approximating to this view have been held by the leading advocates of the cell doctrine, and have beyond question been the inspiration of the development of that doctrine.

But certainly no such conception of the significance of cell structure would longer be held. In spite of the fact that the egg is a single cell, it is impossible to avoid the belief that in some way it contains the starfish. We need not, of course, think of it as containing the structure of a starfish, but we are forced to conclude that in some way its structure is such that it contains the starfish potentially. The relation of its parts and the forces therein are such that, when placed under proper conditions, it develops into a starfish. Another egg placed under identical conditions will develop into a sea urchin, and another into an oyster. If these three eggs have the power of developing into three different animals under identical conditions, it is evident that they must have corresponding differences in spite of the fact that each is a single cell. Each must in some way contain its corresponding adult. In other words, the organization must be within the cells, and hence not simply produced by the associations of cells.

Over this subject there has been a deal of puzzling and not a little experimentation. The presence of some sort of organization in the egg is clear—but what is meant by this statement is not quite so clear. Is this adult organization in the whole egg or only in its nucleus, and especially in the chromosomes which, as we have seen, contain the hereditary traits? When the egg begins to divide does each of the first two cells still contain potentially the organization of the whole adult, or only one half of it? Is the development of the egg simply the unfolding of some structure already present; or is the structure constantly developing into more and more complicated conditions owing to the bringing of its parts into new relations? To answer these questions experimenters have been engaged in dividing developing eggs into pieces to determine what powers are still possessed by the fragments. The results of such experiments are as yet rather conflicting, but it is evident enough from them that we can no longer look upon the egg cell as a simple undifferentiated cell. In some way it already contains the characters of the adult, and when we remember that the characters of the adult which are to be developed from the egg are already determined, even to many minute details—such, for instance, as the inheritance of a congenital mark—it becomes evident that the egg is a body of extraordinary complexity. And yet the egg is nothing more than a single cell agreeing with other cells in all its general characters. It is clear, then, that we must look upon organization as something superior to cells and something existing within them, or at least within the egg cell, and controlling its development. We are forced to believe, further, that there may be as important differences between two cells as there are between two adult animals or plants. In some way there must be concealed within the two cells which constitute the egg of the starfish and the man differences which correspond to the differences between the starfish and the man. Organization, in other words, is superior to cell structure, and the cell itself is an organization of smaller units.

As the result of these various considerations there has been, in recent years, something of a reaction against the cell doctrine as formerly held. While the study of cells is still regarded as the key to the interpretation of life phenomena, biologists are seeing more and more clearly that they must look deeper than simple cell structure for their explanation of the life processes. While the study of cells has thrown an immense amount of light upon life, we seem hardly nearer the centre of the problem than we were before the beginning of the series of discoveries inaugurated by the formulation of the doctrine of protoplasm.

Fundamental Vital Activities as Located in Cells.—We are now in position to ask whether our knowledge of cells has aided us in finding an explanation of the fundamental vital actions to which, as we have seen, life processes are to be reduced. The four properties of irritability, contractibility, assimilation, and reproduction, belong to these vital units— the cells, and it is these properties which we are trying to trace to their source as a foundation of vital activity.

We may first ask whether we have any facts which indicate that any special parts of the cell are associated with any of these fundamental activities. The first fact that stands out clearly is that the nucleus is connected most intimately with the process of reproduction and especially with heredity. This has long been believed, but has now been clearly demonstrated by the experiments of cutting into fragments the cell bodies of unicellular animals. As already noticed, those pieces which possess a nucleus are able to continue their life and reproduce themselves, while those without a nucleus are incapable of reproduction. With greater force still is the fact shown by the process of fertilization of the egg. The egg is very large and the male reproductive cell is very small, and the amount of material which the offspring derives from its mother is very great compared with that which it derives from its father. But the child inherits equally from father and mother, and hence we must find the hereditary traits handed down in some element which the offspring obtains equally from father and mother. As we have seen (Figs. 34-44), the only element which answers this demand is the nucleus, and more particularly the chromosomes of the nucleus. Clearly enough, then, we must look upon the nucleus as the special agent in reproduction of cells.

Again, we have apparently conclusive evidence that the nucleus controls that part of the assimilative process which we have spoken of as the constructive processes. The metabolic processes of life are both constructive and destructive. By the former, the material taken into the cell in the form of food is built up into cell tissue, such as linin, microsomes, etc., and, by the latter, these products are to a greater or less extent broken to pieces again to liberate their energy, and thus give rise to the activities of the cell. If the destructive processes were to go on alone the organism might continue to manifest its life activities for a time until it had exhausted the products stored up in its body for such purposes, but it would die from the lack of more material for destruction. Life is not complete without both processes. Now, in the life of the cell we may apparently attribute the destructive processes to the cell substance and the constructive processes to the nucleus. In a cell which has been cut into fragments those pieces without a nucleus continue to show the ordinary activities of life for a time, but they do not live very long (Fig. 25). The fragment is unable to assimilate its food sufficiently to build up more material. So long as it still retains within itself a sufficiency of already formed tissue for its destructive metabolism, it can continue to move around actively and behave like a complete cell, but eventually it dies from starvation. On the other hand, those fragments which retain a piece of the nucleus, even though they have only a small portion of the cell substance, feed, assimilate, and grow; in other words, they carry on not only the destructive but also the constructive changes. Plainly, this means that the nucleus controls the constructive processes, although it does not necessarily mean that the cell substance has no share in these constructive processes. Without the nucleus the cell is unable to perform those processes, while it is able to carry on the destructive processes readily enough. The nucleus controls, though it may not entirely carry on, the constructive metabolism.

It is equally clear that the cell substance is the seat of most of the destructive processes which constitute vital action. The cell substance is irritable, and is endowed with the power of contractility. Cell fragments without nucleii are sensitive enough, and can move around as readily as normal cells. Moreover, the various fibres which surround the centrosomes in cell division and whose contractions and expansions, as we have seen, pull the chromosomes apart in cell division, are parts of the cell substance. All of these are the results of destructive metabolism, and we must, therefore, conclude that destructive processes are seated in the cell substance.

The centrosome is too problematical as yet for much comment. It appears to be a piece of the machinery for bringing about cell division, but beyond this it is not safe to make any statements.

In brief, then, the cell body is a machine for carrying on destructive chemical changes, and liberating from the compounds thus broken to pieces their inclosed energy, which is at once converted into motion or heat or some other form of active energy. This chemical destruction is, however, possible only after the chemical compounds have become a part of the cell. The cell, therefore, possesses a nucleus which has the power of enabling it to assimilate its food—that is, to convert it into its own substance. The nucleus further contains a marvellous material—chromatin—which in someway exercises a controlling influence in its life and is handed down from one generation to another by continuous descent. Lastly, the cell has the centrosome, which brings about cell division in such a manner that this chromatin material is divided equally among the subsequent descendants, and thus insures that the daughter cells shall all be equivalent to each other and to the mother cell.

We must therefore look upon the organic cell as a little engine with admirably adapted parts. Within this engine chemical activity is excited. The fuel supplied to the engine is combined by chemical forces with the oxygen of the air. The vigour of the oxidation is partly dependent upon temperature, just as it is in any other oxidation process, and is of course dependent upon the presence of fuel to be oxidized, and air to furnish the oxygen. Unless the fuel is supplied and the air has free access to it, the machine stops, the cell dies. The energy liberated in this machine is converted into motion or some other form. We do not indeed understand the construction of the machine well enough to explain the exact mechanism by which this conversion takes place, but that there is such a mechanism can not be doubted, and the structure of the cell is certainly complex enough to give plenty of room for it. The irritability of the cell is easily understood; for, since it is made of very unstable chemical compounds, any slight disturbance or stimulation on one part will tend to upset its chemical stability and produce reaction; and this is what is meant by irritability.

Or, again, we may look upon the cell as a little chemical laboratory, where chemical changes are constantly occurring. These changes we do not indeed understand, but they are undoubtedly chemical changes. The result is that some compounds are pulled to pieces and part of the fragments liberated or excreted, while other parts are retained and built into other more complex compounds. The compounds thus manufactured are retained in the cell body, and it grows in bulk. This continues until the cell becomes too big, and then it divides.

If a machine is broken it ceases to carry on its proper duties, and if the parts are badly broken it is ruined. So with the cell. If it is broken by any means, mechanical, thermal, or otherwise, it ceases to run—we say it dies. It has within itself great power of repairing injury, and therefore it does not cease to act until the injury is so great as to be beyond repair. Thus it only stops its motion when the machinery has become so badly injured as to be beyond hope of repair, and hence the cell, after once ceasing its action, can never resume it again.

There are, of course, other functions of living things besides the few simple ones which we have considered. But these are the fundamental ones; and if we can reduce them to an intelligible explanation, we may feel that we have really grasped the essence of life. If we understand how the cell can move and grow and reproduce itself, we may rest assured that the other phenomena of life follow as a natural consequence. If, therefore, we have obtained an understanding of these fundamental vital phenomena, we have accomplished our object of comprehending the life phenomena in our chemical and mechanical laws.

But have we thus reduced these fundamental phenomena to an intelligible explanation? It must be acknowledged that we have not. We have reduced them to the action of chemical forces acting in a machine. But the machine itself is unintelligible. The organic cell is no more intelligible to us than is the body as a whole. The chemical understanding which we thought we had a few years ago in protoplasm has failed us, and nothing has taken its place We have no conception of what may be the primitive life substance. All we can say is that this most marvellous of all natural phenomena occurs only within that peculiar piece of machinery which we call the cell, and that it is the result of the action of physical forces in that machine. How the machine acts, or even the structure of the machine, we are as far from understanding as we were fifty years ago. The solution has retreated before us even faster than we have advanced toward it.

Summary.—We may now notice in a brief summary the position which we have reached. In our attempt to explain the living organism on the principle of the machine, we are very successful so far as secondary problems are concerned. Digestion, circulation, respiration, and motion are readily solved upon chemical and mechanical principles. Even the phenomena of the nervous system are, in a measure, capable of comprehension within a mechanical formula, leaving out of account the purely mental phenomena which certainly have not been touched by the investigation. All of these phenomena are reducible to a few simple fundamental activities, and these fundamental activities we find manifested by simple bits of living matter unincumbered by the complicated machinery of organisms. With the few fundamental properties of these bits of organic matter we can construct the complicated life of the higher organism. When we come, however, to study these simple bits of matter, they prove to be anything but simple bits of matter. They, too, are pieces of complicated mechanism whose action we do not even hope to understand. That their action is dependent upon their machinery is evident enough from the simple description of cell activity which we have noticed. That these fundamental vital properties are to be explained as the result of chemical and mechanical forces acting through this machinery, can not be doubted. But how this occurs or what constitutes the guiding force which corresponds to the engineer of the machine, we do not know.

Thus our mechanical explanation of the living machine lacks a foundation. We can understand tolerably well the building of the superstructure, but the foundation stones upon which that structure is built are unintelligible to us. The running of the living machine is thus only in part understood. The living organism is a machine or, it is better to say, it is a series of machines one within the other. As a whole it is a machine, and its parts are separate machines. Each part is further made up of still smaller machines until we reach the realm of the microscope. Here still we find the same story. Even the parts formerly called units, prove to be machines, and when we recognize the complexity of these cells and their marvellous activities, we are ready to believe that we may find still further machines within. And thus vital activity is reduced to a complex of machines, all acting in harmony with each other to produce together the one result—life.

PART II.
THE BUILDING OF THE LIVING MACHINE.

CHAPTER III.
THE FACTORS CONCERNED IN THE BUILDING OF THE LIVING MACHINE.

Having now outlined the results of our study into the mechanism of the living machine, we turn our attention next to the more difficult problem of the method by which this machine was built. From the facts which we have been considering in the last two chapters it is evident that the problem we have before us is a mechanical rather than a chemical one. Of course, chemical forces lie at the bottom of vital activity, and we must look upon the force of chemical affinity as the fundamental power to which the problems must be referred. But a chemical explanation will evidently not suffice for our purpose; for we have absolutely no reason for believing that the phenomena of life can occur as the results of the chemical properties of any compound, however complex. The simplest known form of matter which manifests life is a machine, and the problem of the origin of life must be of the origin of that machine. Are there any forces in nature which are of a sort as to enable us to use them to explain the building of machines? Plants and animals are the only machines which nature has produced. They are the only instances in nature of a structure built with its parts harmoniously adjusted to each other to the performance of certain ends. All other machines with which we are acquainted were made by man, and in making them intelligence came in to adapt the parts to each other. But in the living organism is a similarly adapted machine made by natural means rather than artificial. How were they built? Does nature, apart from human intelligence, possess forces which can achieve such results?

Here again we must attack the problem from what seems to be the wrong end. Apparently it would be simpler to discover the method of the manufacture of the simplest machine rather than the more complex ones. But this has proved contrary to the fact. Perhaps the chief reason is that the simplest living machine is the cell whose study must always involve the use of the microscope, and for this reason is more difficult. Perhaps it is because the problem is really a more difficult one than to explain the building of the more complex machines out of the simpler ones. At all events, the last fifty years have told us much of the method of the building of the complex machines out of the simpler ones, while we have as yet not even a hint as to the solution of the building of the simplest machine from the inanimate world. Our attention must, therefore, be first directed to the method by which nature has constructed the complex machines which we find filling the world to-day in the form of animals and plants.

History of the Living Machine.—In the first place, we must notice that these machines have not been fashioned suddenly or rapidly, but have been the result of a very slow growth. They have had a history extending very far back into the past for a period of years which we can only indefinitely estimate, but certainly reaching into the millions. As we look over this past history in the light of our present knowledge we see that whatever have been the forces which have been concerned in the construction of these machines they have acted very slowly. It has taken centuries, and, indeed, thousands of years, to take the successive steps which have been necessary in this construction. Secondly, we notice that the machines have been built up step by step, one feature being added to another with the slowly progressing ages. Thirdly, we notice that in one respect this construction of the living machine by nature's processes has been different from our ordinary method of building machines. Our method of building puts the parts gradually into place in such a way that until the machine is finished it is incapable of performing its functions. The half-built engine is as useless and as powerless as so much crude iron. Its power of action only appears after the last part is fitted into place and the machine finished. But nature's process in machine building is different. Every step in the process, so far as we can trace it at least, has produced a complete machine. So far back as we can follow this history we find that at every point the machine was so complete as to be always endowed with motion and life activity. Nature's method has been to take simpler types of machines and slowly change them into more complicated ones without at any moment impairing their vigour. It is something as if the steam engine of Watt should be slowly changed by adding piece after piece until there was finally produced the modern quadruple expansion engine, but all this change being made upon the original engine without once stopping its motion.

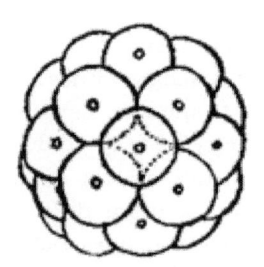

FIG. 45.
A group of cells resulting from division, representing the first step in machine making.

This gradual construction of the living machines has been called Organic Evolution, or the Theory of Descent. It will be necessary for us, in order to comprehend the problem which we have before us, to briefly outline the course of this evolution. Our starting point in this history must be the cell, for such is the earliest and simplest form of living thing of which we have any trace. This cell is, of course, already a machine, and we must presently return to the problem of its origin. At present we will assume this cell as a starting point endowed with its fundamental vital powers. It was sensitive, it could feel, grow, and reproduce itself. From such a simple machine, thus endowed, the history has been something as follows: In reproducing itself this machine, as we have already seen, simply divided itself into two halves, each like the other. At first all the parts thus arising separated from each other and remained independent. But so long as this habit continued there could be little advance. After a time some of the cells failed to separate after division, but remained clinging together (Fig. 45). The cells of such a mass must have been at first all alike; but, after a little, differences began to appear among them. Those on the outside of the mass were differently affected by their surroundings from those in the interior, and soon the cells began to share among themselves the different duties of life. The cells on the outside were better situated for protection and capturing food, while those on the inside could not readily seize food for themselves, and took upon themselves the duty of digesting the food which was handed to them by the outer cells. Each of these sets of cells could now carry on its own special duties to better advantage, since it was freed from other duties, and thus the whole mass of cells was better served than when each cell tried to do everything for itself. This was the first step in the building of the machine out of the active cells (Fig. 46). From such a starting point the subsequent history has been ever based upon the same principle. There has been a constant separation of the different functions of life among groups of cells, and as the history went on this division of labor among the different parts became greater and greater. Group after group of cells were set apart for one special duty after another, and the result was a larger and ever more complicated mass of cells, with a greater and greater differentiation among them. In this building of the machine there was no time when the machine was not active. At all points the machine was alive and functional, but each step made the total function of the machine a little more accurately performed, and hence raised somewhat the totality of life powers. This parcelling out of the different duties of life to groups of cells continued age after age, each step being a little advance over the last, until the result has been the living machine as we know it in its highest form, with its numerous organs, all interrelated in such a way as to form a harmoniously acting whole.

FIG. 40.

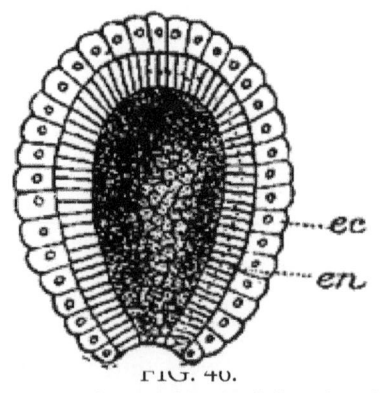

A later step in machinebuilding in which the outer cells have acquired different form and function from the inner cells: ec, the outer cells, whose duties are protective; en, the inner cells engaged in digesting food.

But a second principle in this growth of the machine was needed to produce the variety which is found in nature. As the different cells in the multicellular mass became associated into groups for different duties, the method of such division of labor was not alike in all machines. A city in China and one in America are alike made up of individuals, and the fundamental needs of the Chinaman and the American are alike. But differences in industrial and political conditions have produced different combinations and associations, so that Pekin is wonderfully unlike New York. So in these early developing machines, quite a variety of method of organization was adopted by the different groups. Now as soon as any special type of organization was adopted by any animal or plant, the principle of heredity transmitted the same kind of organization to its descendants, and there thus arose lines of descent differing from each other, each line having its own method of organization. As we follow the history of each line the same thing is repeated. We find that the representatives of each line again separate into groups, each of which has acquired some new type of organization, and there has thus been a constant divergence of these lines of descent in an indefinite number of directions. The members of the different lines of descent all show a fundamental likeness with each other since they retain the fundamental characters of their common ancestor, but they show also the differences which they have themselves acquired. And thus the process is repeated over and over again. This history of the growth of these different machines has thus been one of divergence from common centres, and is to be diagrammatically expressed after the fashion of a branching tree. The end of each branch represents the highest state of perfection to which each line has been carried.

One other point in this history must be noted. As the development of the complication of the machine progressed the possibility of further progress has been constantly narrowed. When the history of these machines began as a simple mass of cells, there was a possibility of an almost endless variety of methods of organization. But as a distinct type of organization was adopted by one and another line of descendants all subsequent productions were limited through the law of heredity to the general line of organization adopted by their ancestors. With each age the further growth of such machines must consist in the further development in the perfection of its parts, and not in the adoption of any new system of organization. Hence it is that the history of the living machine has shown a tendency toward development along a few well-marked lines, and although this complication becomes greater, we still see the same fundamental scheme of organization running through the whole. As the ages have progressed the machines have become more perfect in the adjustment of their parts, i.e., they have become more perfect machines, but the history has been simply that of perfecting the early machines rather than the production of new types.

Evidence for this History.—As just outlined, we see that the living machines have been gradually brought into their present condition by a process which has been called organic evolution. But we must pause for a moment to ask what is our evidence that such has been the history of the living machine. The whole possibility of understanding living nature depends upon our accepting this history and finding an explanation of it. At the outset we have the question of fact, and we must notice the grounds upon which we stand in assuming this history to be as outlined.

This problem is the one which has occupied such a prominent place in the scientific world during the last forty years, and which has contributed so largely toward making modern biology such a different subject from the earlier studies of natural history. It is simply the evidence for organic evolution, or the theory of descent. The subject has for forty years been thoroughly sifted and tested by every conceivable sort of test. As a result of the interest in the question there has been disclosed an immense mass of evidence, relevant and irrelevant. As the evidence has accumulated it has become more and more evident that the evolution theory must be recognized as the only one which is in accord with the facts, and the outcome has been a practical unanimity among thinkers that the theory of descent must be the foundation of our further study. The evidence which has forced this conclusion upon scientists we must stop for a moment to consider, since it bears very directly upon the subject we are studying.

Historical.—The first source of evidence is naturally a historical one. This long history of the construction of the living machine has left its record in the rocks which form the earth's surface. During this long period the rocks of the earth's crust have been deposited, and in these rocks have been left samples of many of the steps in this history of machine building. The history can be traced by the study of these samples just as the history of any machine might be traced from a study of the models in a patent office. One might very easily trace, with most strict accuracy and minute detail, the history of the printing machine from the models which are preserved in the patent offices and elsewhere. So is it with the history of the living machine. To be sure, the history is rather incomplete and at times difficult to read. Many a period in the development has left no samples for our inspection and must be interpreted in our history between what went before and what comes after. Many of the machines, especially the early ones, were made of such fragile material that they could not be preserved in the rocks. In many a case, too, the rocks in which the specimens were deposited have been subjected to such a variety of heatings and pressures, that they have been twisted out of shape and even crushed out of recognizable form. But in spite of this the record is showing itself more complete each year. Our paleontologists are opening layer after layer of these rocks, and thus examining each year new pages in nature's history. The more recent epochs in the history have been already read with almost historic accuracy. From them we have learned in great detail how the finishing touches were given to these machines, and are able to trace with accuracy how the somewhat more generalized forms of earlier days were changed to produce our modern animals.

This fossil record has given us our best knowledge of the course by which the present living world has been brought into its existing condition. But its accuracy is largely confined to the recent periods. Of the very early history fossils tell us little or nothing. All the early rocks, which we may believe were formed during the period when the first steps in this machine building were taken, have been so changed by heat and pressure that whatever specimens they may have originally contained have been crushed out of shape. Furthermore, the earliest organisms had no hard skeletons, and it was not until living beings had developed far enough to have hard parts that it was possible for them to leave traces of themselves in the rocks. Hence, so far as concerns this earliest history, we can get no record of it in the rocks.

Embryological.—But here comes in another source of evidence which helps to fill up the gap. In its development every animal to-day begins as an egg. This is a simple cell, and the animal goes through a series of changes which eventually lead to the adult. Now these changes appear for the most part to be parallel to the changes through which the earlier forms of life passed in their development from the simple to the more complicated forms. Where it is possible to follow the history of the groups of animals from their fossil remains and compare it with the history of the individual animal as it progresses from the egg to the adult, there is found a very decided parallelism. This parallelism between embryology and past history has been of great service in helping us toward the history of the past. At one time it was believed that it was the key which would unlock all doors, and for a decade biologists eagerly pursued embryology with the expectation that it would solve all problems in connection with the history of animals. The result has been somewhat disappointing. Embryology has, it is true, been of the utmost service in showing relationships of forms to each other, and in thus revealing past history. But while this record is a valuable one, it is a record which has unfortunately been subject to such modifying conditions that in many cases its original meaning has been entirely obliterated and it has become worthless as a historical record. These imperfections in regard to the record were early seen after the attention of biologists was seriously turned to the study of embryology, but it was expected that it would be possible to correct them and discover the true meaning underlying the more apparent one. Indeed, in many cases this has been found possible. But many of the modifications are so profound as to render it impossible to untangle them and discover the true meaning. As a result the biologist to-day is showing less confidence in embryology, and is turning his attention in different directions as more promising of results in the line desired.

But although the teachings of embryology have failed to realize the great hopes that were placed upon them, their assistance in the formulation of this history of the machine has been of extreme value. Many a bit of obscurity has been cleared up when the embryology of puzzling animals has been studied. Many a relationship has been made clear, and this is simply another way of saying that a portion of this history of life has been read. This aid of embryology has been particularly valuable in just that part of the history where the evidence from the study of fossils is wanting. The study of fossils, as we have seen, gives little or no data concerning the early history of living machines; and it is just here that embryology has proved to be of the most value. It is a source of evidence that has told us of most of the steps in the progress from the single-celled animal to the multicellular organisms, and gives us the clearest idea of the fundamental principles which have been concerned in the evolution of life and the construction of the complicated machine out of the simple bit of protoplasm. In spite of its limits, therefore, embryology has contributed a large quota of the evidence which we have of the evolution of life.

Anatomical.—A third source of this history is obtained from the facts of comparative anatomy. The essential feature of this subject is the fact that animals and plants show relationships. This fact is one of the most patent and yet one of the most suggestive facts of biology. It has been recognized from the very beginning of the study of animals and plants. One cannot be even the most superficial observer without seeing that certain forms show great likeness to each other while others are much more unlike. The grouping of animals and plants into orders, genera, and species is dependent upon this relationship. If two forms are alike in everything except some slight detail, they are commonly placed in the same genus but in different species, while if they show a greater unlikeness they may be placed in separate genera. By thus grouping together forms according to their resemblance the animal and vegetable kingdoms are classified into groups subordinate to groups. The principle of relationship, i.e., fundamental similarity of structure, runs through the whole animal and vegetable kingdom. Even the animals most unlike each other show certain points of similarity which indicates a relationship, although of course a distant one.

The fact of such a relationship is too patent to demand more words, but its significance needs to be pointed out. When we speak of relationship among men we always mean historical connection. Two brothers are closely related because they have sprung from common parents, while two cousins are less closely related because their common point of origin was farther back in time. More widely we speak of the relationship of the Indo-European races, meaning thereby that back in the history of man these races had a common point of origin. We never speak of any real relation of objects unless thereby we mean to imply historical connection. We are therefore justified in interpreting the manifest relationships of organisms as pointing to history. Particularly are we justified in this conclusion when we find that the relationships which we draw between the types of life now in existence run parallel to the history of these types as revealed to us by fossils and at the same time disclosed by the study of embryology.

This subject of comparative anatomy includes a consideration of what is called homology, and perhaps a concrete example may be instructive both in illustration and as suggesting the course which nature adopts in constructing her machines. We speak of a monkey's arm and a bird's wing as homologous, although they are wonderfully different in appearance and adapted to different duties. They are called homologous because they have similar parts in similar relations. This can be seen in Figs. 47 and 48, where it will be seen that each has the same bones, although in the bird's wing some of the bones have been fused together and others lost. Their similarity points to a relationship, but their dissimilarity tells us that the relationship is a distant one, and that their common point of origin must have been quite far back in history. Now if we follow back the history of these two kinds of appendages, as shown to us by fossils, we find them approaching a common point. The arm can readily be traced to a walking appendage, while the bird's wing, by means of some interesting connecting links, can in a similar way be traced to an appendage with its five fingers all free and used for walking. Fig. 49 shows one of these connecting links representing the earliest type of bird, where the fingers and bones of the arm were still distinct, and yet the whole formed a true wing. Thus we see that the common point of origin which is suggested by the likenesses between an arm and a wing is no mere imaginary one, for the fossil record has shown us the path leading to that point of origin. The whole tells us further that nature's method of producing a grasping or flying organ was here, not to build a new organ, but to take one that had hitherto been used for other purposes, and by slow changes modify its form and function until it was adapted to new duties.

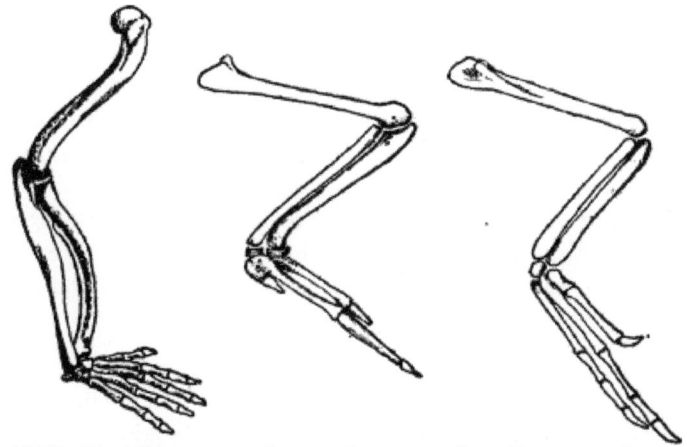

FIG. 47.—The arm of a monkey, a prehensile appendage.
FIG. 48.—The arm of a bird, a flying appendage. In life covered with feathers.
FIG. 49.—The arm of an ancient half-bird half-reptile animal. In life covered with feathers and serving as a wing.

Significance of these Sources of History.—The real force of these sources of evidence comes to us only when we compare them with each other. They agree in a most remarkable fashion. The history as disclosed by fossils and that told by embryology agree with each other, and these are in close harmony with the history as it can be read from comparative anatomy. If archæologists were to find, in different countries and entirely unconnected with each other two or more different records of a lost nation, the belief in the actual existence of that nation would be irresistible. When researches at

Nineveh, for example, unearth tablets which give the history of ancient nations, and when it proves that among the nations thus mentioned are some with the same names and having the same facts of history as those mentioned in the Bible, it is absolutely impossible to avoid the conclusion that such a nation with such a history did actually exist. Two independent sources of record could not be false in regard to such a matter as this.

Now, our sources of evidence for this history of the living machine prove to be of exactly this kind. We have three independent sources of evidence which are so entirely different from each other that there is almost no likeness between them. One is written in the rocks, one in bone and muscle, while the third is recorded in the evanescent and changing pages of embryology and metamorphosis. Yet each tells the same story. Each tells of a history of this machine from simple forms to more complex. Each tells of its greater and greater differentiation of labour and structure as the periods of time passed. Each tells of a growing complexity and an increasing perfection of the organisms as successive periods pass. Each tells us of common points of origin and divergence from these points. Each tells us how the more complicated forms have arisen as the results of changes in and modifications of the simpler forms. Each shows us how the individual parts of the organisms have been enlarged or diminished or changed in shape to adapt them to new duties. Each, in short, tells the same story of the gradual construction of the living machine by slow steps and through long ages of time. When these three sources of history so accurately agree with each other, it is as impossible to disbelieve in the existence of such history as it is to disbelieve in the existence of the ancient Hittite nation, after its history has been told to us by two different sources of record.

Now all this is very germane to our subject. We are trying to learn how this living machine, with its wonderful capabilities, was built. The history which we have outlined is undoubtedly the history of the building of this machine, and the knowledge that these complicated machines have been produced as the result of slow growth is of the utmost importance to us. This knowledge gives us at the very start some idea of the nature of the forces which have been at work. It tells us that in searching for these forces we must look for those which have been acting constantly. We must look for forces which produce their effects not by sudden additions to the complication of the machine. They must be constant forces whose effect at any one time is comparatively slight, but whose total effect is to increase the complexity of the machine. They must be forces which produce new types through the modification of the old ones. We must look for forces which do not adapt the machine for its future, but only for its present need. Each step in the history has been a complete animal with its own fully developed powers. We are not to expect to find forces which planned the perfect machine from the start, nor forces which were engaged in constructing parts for future use. Each step in the building of the machine was taken for the good of the machine at the particular moment, and the forces which we are to look for must therefore be only such as can adapt the organisms for its present needs. In other words, nothing has been produced in this machine for the purpose of being developed later into something of value, but all parts that have been produced are of value at the time of their appearance. We must, in short, look for forces constantly in action and always tending in the same direction of greater complexity of structure.

Is it possible to discover these forces and comprehend their action? Before the modern development of evolution this question would unhesitatingly have been answered in the negative. To-day, under the influence of the descent theory, stimulated, in the first place, by Darwin, the question will be answered by many with equal promptness in the affirmative. At all events, we have learned in the last forty years to recognize some of the factors which have been at work in the construction of this machine. We must turn, therefore, to the consideration of these factors.

Forces at Work in the Building of the Living Machine.—There are three primary factors which lie at the bottom of the whole process. They are—

1. Reproduction, which preserves type from generation to generation.

2. Variation, which modifies type from generation to generation.

3. Heredity, which transmits characters from generation to generation.

Each must be considered by itself.

Reproduction.—Reproduction is the primary factor in this process of machine building, heredity and variation being simply phases of reproduction. The living machine has developed by natural processes, all other machines by artificial methods. Reproduction is the one essential point of difference between the living machine and the others which has made their construction by natural processes a possibility. What, then, is reproduction? Reproduction is in all cases at the bottom simple division. Whether we consider the plant that multiplies by buds or the unicellular animal that simply divides into two equal parts, or the larger animal that multiplies by eggs, we find that in all cases the fundamental feature of the process is division. In all cases the organism divides into two or more parts, each of which becomes in time like the original. Moreover, when we trace this division further we find that in all cases it is to be referred back to the division of the cell, such as we have described in a previous chapter. The egg is a single cell which has come from the parent by the division of one of the cells in the body of the parent. A bud is simply a mass of cells which have all arisen from the parent cells by division. The foundation of reproduction is thus in all cases cell division. Now, this process of division is dependent upon the properties of the cell. Firstly, it is a result of the assimilative powers of the cell, for only through assimilation can the cell increase in size, and only as it increases in size can it gain sustenance for cell division. Secondly, it is dependent, as we have seen, upon the mechanism of the cell body, and especially the nucleus and centrosome. These structures regulate the cell division, and hence the reproduction of all animals and plants. We can not, therefore, find any explanation of reproduction until we have explained the mechanism of the cell. The fundamental feature, of nature's machine building is thus based upon the machinery of the nucleus and centrosome of the organic cell.

Aside from the simple fact that it preserves the race, the most important feature connected with this reproduction is its wonderful fruitfulness. Since it results from division, it always tends to increase the offspring in geometrical ratio. In the simplest case, that of the unicellular animals, the cell divides, giving rise to two animals, each of which divides again, producing four, and these again, giving eight, etc. The rapidity of this multiplication is sometimes inconceivable. It depends, of course, upon the interval of time between the successive divisions, but among the lower organisms this interval is sometimes not more than half an hour, the result of which is that a single individual could give rise in the course of twenty-four hours to sixteen million offspring. This is doubtless an extreme case, but among all the lower animals the rate is very great. Among larger animals the process is more complicated; but here, too, there is the same tendency to geometrical progression, although the intervals between the successive reproductions may be quite long and irregular. But it is always so great that if allowed to progress unhindered at its normal rate the offspring would, in a few years, become so numerous as to crowd other life out of existence. Even the slow-breeding elephant would, if allowed to breed unhindered for seven hundred and fifty years, produce nineteen million offspring—a rate of increase plainly incompatible with the continued existence of other animals.

Here, then, we have the foundation of nature's method of building animals and plants of the higher classes. In the machinery of the cell she has a power of reproduction which produces an increase in geometrical ratio far beyond the possibility for the surface of the earth to maintain.

Heredity.—The offspring which arise by these processes of division are like each other, and like the parent from which they sprung. This is the essence of what is called heredity. Its significance in the process of machine building is evident at once. It is the conserving force which preserves the forms already produced and makes it possible for each generation to build upon the structures of the earlier ones. Without it each generation would have to begin anew at the beginning, and nothing could be accomplished. But since this principle brings each individual to the same place where its parents stand, and thus always builds the offspring into a machine like the parent, it makes it possible for the successive generations to advance. Heredity is thus like the power of memory, or better still, like the invention of printing in the development of civilization. It is a record of past achievements. By means of printing each age is enabled to benefit by the discoveries of the previous age, and without it the development of civilization would be impossible. In the same way heredity enables each generation to benefit by the achievements of its ancestors in the process of machine building, and thus to devote its own energies to advancement.

The fact of heredity is patent enough. It has been always clearly recognized that the child has the characters of its parents, and this belief is so well attested as to need no proof. It is still a question as to just what characters may be inherited, and what influences may affect the inheritance. There are plenty of puzzling problems connected with heredity, but the fact of heredity is one of the foundation stones of biological science. Upon it must be built all theories which look toward the explanation of the origin of the living machine.

This factor of heredity again we must trace back to the machinery of the cell. We have seen in the previous pages evidence for the wonderful nature of the chromosomes of the cells. We can not pretend to understand them, but they must be extraordinarily complex. We have seen proof that these chromosomes are probably the physical basis of heredity, since they are the only parts of each parent which are handed down to subsequent generations. With these various facts of cell division and cell fertilization in mind, we can reach a very simple explanation of fundamental features of heredity. The following is an outline of the most widely accepted view of the hereditary process.

Recognizing that the chromosomes are the physical basis of hereditary transmission, we can picture to ourselves the transmission of hereditary characters something as follows: As we have seen, the fertilized egg contains an equal number of chromosomes from each parent (Fig. 42). Now when this fertilized cell divides, each of the rods splits lengthwise, half of each entering each of the two cells arising from the cell division. From this method of division of the chromosomes it follows that the daughter cells would be equivalent to each other and equivalent also to the undivided egg. If the original chromosomes contained potentially all the hereditary traits handed down from parent to child, the chromosomes of each daughter cell will contain similar hereditary traits. If, therefore, the original fertilized egg possessed the power of developing into an adult like the parent, each of the daughter cells should likewise possess the power of developing into a similar adult. And thus each cell which arises as the result of such division should possess similar characters so long as this method of division continues. But after a little in the development of the egg a differentiation among the daughter cells arises. They begin to acquire different shapes and different functions. This we can only believe to be the result of a differentiation in their chromatin material. In the cell division the chromosomes no longer split into equivalent halves, but some characters are portioned off to some cells and others to other cells. Those cells which are to carry on digestive functions when they are formed receive chromatin material which especially controls them in the performance of this digestive function, while those which are to produce sensory organs receive a different portion of the chromatin material. Thus the adult individual is built up as the cells receive different portions of this hereditary substance contained in the original chromosomes. The original chromosomes contained all hereditary characters, but as development proceeds these are gradually portioned out among the daughter cells until the adult is formed.

From this method of division it will be seen that each cell of the adult does not contain all the characters concealed in the original chromosomes of the egg, although each contains a part which may have been derived from each parent. It is thought, however, that a part of the original chromatin material does not thus become differentiated, but remains entirely unchanged as the individual is developing. This chromatin material may increase in amount by assimilation, but it remains unchanged during the entire growth of the individual. It thus follows that the adult will contain, along with its differentiated material, a certain amount of the original physical basis of heredity which still retains its original powers. This undifferentiated chromatin material originally possessed powers of producing a new individual, and of course it still possesses these powers, since it has remained dormant without alteration. Further, it will follow that if this dormant undifferentiated chromatin should start into activity and produce a new individual, the new individual thus produced would be identical in all characters with the one which actually did develop from the egg, since both individuals would have come from a bit of the same chromatin. The child would be like the parent. This would be true no matter how much this undifferentiated material should increase in amount by assimilation, so long as it remained unaltered in character, and it hence follows that every individual carries around a certain amount of undifferentiated chromatin material in all respects identical with that from which he developed.

Now whether this undifferentiated germ plasm, as we will now call it, is distributed all over the body, or is collected at certain points, is immaterial to our purpose. It is certain that portions of it find their way into the reproductive organs of the animal or plant. Thus we see that part of the chromatin material in the egg of the first generation develops into the second generation, while another part of it remains dormant in that second generation, eventually becoming the chromatin of its eggs and spermatozoa. Thus each egg of the second generation receives chromosomes which have come directly from the first generation, and thus it will follow that each of these eggs will have identical properties with the egg of the first generation. Hence if one of these new eggs develops into an adult it will produce an adult exactly like the second generation, since it contains chromosomes which are absolutely identical with those from which the second generation sprung. There is thus no difficulty in understanding why the second generation will be like the first, and since the process is simply repeated again in the next reproduction, the third generation will be like the second, and so on, generation after generation. A study of the accompanying diagram will make this clear.

In other words, we have here a simple understanding of at least some of the features of heredity. This explanation is that some of the chromatin material or germ plasm is handed down from one generation to another, and is stored temporarily in the nucleii of the reproductive cells. During the life of the individual this germ plasm is capable of increasing in amount without changing its nature, and it thus continues to grow and is handed down from generation to generation, always endowed with the power of developing into a new individual under proper conditions, and of course when it does thus give rise to new individuals they will all be alike. We can thus easily understand why a child is like its parent. It is not because the child can inherit directly from its parent, but rather because both child and parent have come from the unfolding of two bits of the same germ plasm. This fact of the transmission of the hereditary substance from generation to generation is known as the theory of the continuity of germ plasm.

Such appears to be, at least in part, the machinery of heredity. This understanding makes the germ substance perpetual and continuous, and explains why successive generations are alike. It does not explain, indeed, why an individual inherits from its parents, but why it is like its parents. While biologists are still in dispute over many problems connected with heredity, all are agreed to-day that this principle of the continuity of the heredity substance must be the basis of all attempts to understand the machinery of heredity. But plainly this whole process is a function of the cell machinery. While, therefore, the idea of the continuity of germ substance greatly simplifies our problem, we must acknowledge that once more we are thrown back upon the mysteries of the cell. Until we can more fully explain the cell machine we must recognize our inability to solve the fundamental question of why an individual is like its parents.

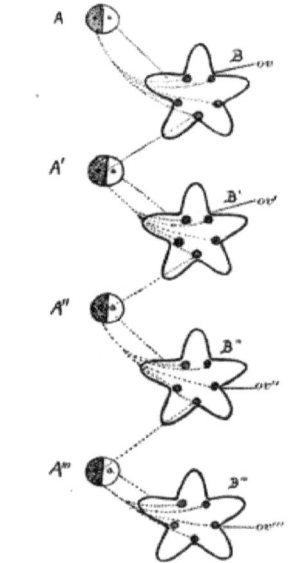

FIG. 50.—Diagram illustrating the principle of heredity.

A represents an egg of a starfish. From one half, the unshaded portion, develops the starfish of the next generation, B. The other is distributed without change in the ovaries, ov, of the individual, B. From these ovaries arises the next egg, A', with its germ plasm. This germ plasm is evidently identical with that in A, since it is merely a bit of the same handed down through the individual, B. In the development of the next generation the process is repeated, and hence B' will be like B, and the third generation of eggs identical with the first and second. The undifferentiated part of the germ plasm is thus simply handed on from one generation to the next.]

But plainly reproduction and heredity, as we have thus far considered them, will be unable to account for the slow modification of the machine; for in accordance with the facts thus far outlined, each generation would be precisely like the last, and there would be no chance for development and change from generation to generation. If the individual is simply the unfolding of the powers possessed by a bit of germ plasm, and if this germ plasm is simply handed on from generation to generation, the successive generations must of necessity be identical. But the living machine has been built by changes in the successive generation, and hence plainly some other factor is needed. This factor is variation.

Variation.—Variation is the principle that produces modification of type. Heredity, as just explained, would make all generations alike. But nothing is more certain than that they are not alike. The fact of variation is patent on every side, for no two individuals are alike. Successive generations differ from each other in one respect or another. Birds vary in the length of their bills or toes; butterflies, in their colours; dogs, in their size and shape and markings; and so on through an endless category. Plants and animals alike throughout nature show variations in the greatest profusion. It is these variations which must furnish us with the foundation of the changes which have gradually built up the living machine.

Of the fact of these variations there is no question, and the matter need not detain us. Every one has had too many experiences to ask for proof. Of the nature of the variations, however, there are some points to be considered which are very germane to our subject. In the first place, we must notice that these variations are of two kinds. There is one class which is born with the individual, so that they are present from the time of birth. In saying that these variations are born with the individual we do not necessarily mean that they are externally apparent at birth. A child may inherit from its parents characters which do not appear till adult life. For example, a child may inherit the colour of its father's hair, but this colour is not apparent at birth. It appears only in later life, but it is none the less an inborn character. In the same way, we may have many inborn variations among individuals which do not make themselves seen until adult life, but which are none the less innate. The offspring of the same parents may show decided differences, although they are put under similar conditions, and such differences are of course inherent in the nature of the individual. Such variations are called congenital variations.

There is, however, a second class of variations which are not born in the individual, but which arise as the result of some conditions affecting its after-life. The most extreme instances of this kind are mutilations. Some men have only one leg because the other has been lost by accident. Here is a variation acquired as the result of circumstances. A blacksmith differs from other members of his race in having exceptionally large arm muscles; but here, again, the large muscles have been produced by use. A European who has lived under a tropical sun has a darkened skin, but this skin has evidently been darkened by the action of the sun, and is quite a different thing from the dark skin of the dark races of men. In such instances we have variations produced in individuals as the result of outside influences acting upon them. They are not inborn, but are secondarily acquired by each individual. We call them acquired variations.

It is not always possible to distinguish between these two types of variation. Frequently a character will be found in regard to which it is impossible to determine whether it is congenital or acquired. If a child is born under the tropical sun, how can we tell whether its dark skin was the result of direct action of the sun on its own skin, or was an inheritance from its dark-skinned parents? We might suppose that this could be answered by taking a similar child, bringing it up away from the tropical sun, and seeing whether his skin remained dark. This would not suffice, however; for if such a child did then develop a white skin, we could not tell but that this lighter-coloured skin had been produced by the direct bleaching effect of the northern climate upon a skin which otherwise would have been dark. In other words, a conclusive answer can not here be given. It is not our purpose, however, to attempt to distinguish between these two kinds of variations, but simply to recognize that they occur.

Our next problem must be to search for an explanation of these variations. With the acquired variations we have no particular trouble, for they are easily explained as due to the direct action of the environment upon animals. One of the fundamental characters of the living protoplasm (using the word now in its widest sense) is its extreme instability. So unstable is it that any disturbing influence will affect it. If two similar unicellular organisms are placed under different conditions they become unlike, since their unstable protoplasm is directly affected by the surrounding conditions. With higher animals the process is naturally a little more complicated; but here, too, they are easily understood as part of the function of the machine. One of the adjustments of the machine is such that when any organ is used more than usual the whole machine reacts in such a way as to send more blood to this special organ. The result is a change in the nutrition of the organ and a corresponding variation in the individual. Thus acquired variations are simply functions of the action of the machine.

Congenital variations, however, can not receive such an explanation. Being born with the individual, they can not be produced by conditions affecting him, but rather to something affecting the germ plasm from which he sprung. The nature of the germ plasm controls the nature of the individual, and congenital variations must consequently be due to its variations. But it is not so easy to see how this germ plasm can undergo variation. The conditions which surround the individual would affect its body, but it is not easy to believe that they would affect the germinal substance. Indeed, it is not easy to see how any external conditions can have influence upon this germinal material if it is not an active part of the body, but is simply stored within it for future use in reproduction. How could any changes in the environment of the individual have any effect upon this dormant material stored within it? But if we are correct in regarding this germ material in the reproductive bodies as the basis of heredity and the guiding force in development, then it follows that the only way in which congenital variations can occur is by some variations in the germ plasm. If a child developed from germ plasm identical with that from which its parents developed, it would inherit identical characters; and if there are any congenital variations from its parents, they must be due to some variations in the germ plasm. In other words, in order to explain congenital variations we must account for variations in the germ plasm.

Now, there are two methods by which we may suppose that these variations in the germ may arise. The first is by the direct influence upon the germ plasm of certain unknown external conditions. The life substance of organisms is always very unstable, and, as we have seen, acquired variations are caused by external influences directly affecting it. Now, the hereditary material is also life substance, and it is plainly a possibility for us to imagine that this germ material is also subject to influences from the conditions surrounding it. That such variations do occur appears to be hardly doubtful, although we do not know what sort of influences can produce them. If the germ plasm is wholly stored within the reproductive gland, it is certainly in a position to be only slightly affected by surrounding conditions which affect the animal. We can readily understand that the use of an organ like the arm will affect it in such a way as to produce changes in its protoplasm, but we can hardly imagine that such use of the arm would produce any change in the hereditary substance which is stored in the reproductive organs. External conditions may thus readily affect the body, but not so readily the germ material. Even if such material is distributed more or less over the body instead of being confined to the reproductive glands, as some believe, the difficulty is hardly lessened. This difficulty of understanding how the germ plasm can be affected by external conditions has led one school of biologists to deny that it is subject to any variation by external conditions, and hence that all modification of the germ plasm must come from some other source. Probably no one, however, holds this position to-day, and it is the general belief that the germ plasm may be to some slight extent modified by external conditions. Of course, if such variations do occur in the germ plasm they will become congenital variations of the next generation, since the next generation is the unfolding of the germ plasm.

The second method by which the variations of germ plasm may arise is apparently of more importance. It is based upon the fact that, with all higher animals and plants at least, each individual has two parents instead of one. In our study of cells we have seen that the machinery of the cell is such that it requires in the ordinary process of reproduction the union of germinal material from two different individuals to produce a cell which can develop into a new individual. As we have seen, the egg gets rid of half its chromosomes in order to receive an equal number from a male parent; and thus the fertilized egg contains chromosomes, and hence hereditary material, from two different individuals. Now, this sexual reproduction occurs very widely in the organic world. Among some of the lowest forms of unicellular organisms it is not known, but in most others some form of such union is universal. Now, here is plainly an abundant opportunity for congenital variations; for it is seen that each individual does not come from germ material identical with that from which either parent came, but from some of this material mixed with a similar amount from a different parent. Now, the two parents are never exactly alike, and hence the germ plasm which each contributes to the offspring will not be exactly alike. The offspring will thus be the result of the unfolding of a bit of germ plasm which will be different from that from which either of its parents developed, and these differences will result in congenital variations. Sexual reproduction thus results in congenital variations; and if congenital variations are necessary for the evolution of the living machine—and we shall soon see reason for believing that they are—we find that sexual reproduction is a device adopted for bringing out such congenital variations.

Inheritance of Variations.—The reason why congenital variations are needed for the evolution of the living machine is clear enough. Evanescent variations can have no effect upon this machine, for they would disappear with the individual in which they appeared. In order that they should have any influence in the process of machine building they must be permanent ones; or, in other words, they must be inherited from generation to generation. Only as such variations are transmitted by heredity can they be added to the structure of the developing machine. Therefore we must ask whether the variations are inherited.

In regard to the congenital variations there can be no difficulty. The very fact that they are congenital shows us that they have been produced by variations in the germ plasm, and as such they must be transmitted, not only to the next generation, but to all following generations, until the germ plasm becomes again modified. This germ plasm is handed on from generation to generation with all its variations, and hence the variations will be added permanently to the machine. Congenital variations are thus a means for permanently modifying the organism, and by their agency must we in large measure believe that evolution through the ages has taken place.

With the acquired variations the matter stands quite differently. We can readily understand how influences surrounding an animal may affect its organs. The increase in the size of the muscles of the blacksmith's arm by use we understand readily enough. But with our understanding of the machinery of heredity we can not see how such an effect can extend to the next generation. It is only the organ directly affected that is modified by external conditions. Acquired variations will appear in the part of the body influenced by the changed conditions. But the germ plasm within the reproductive glands is not, so far as we can see, subject to the influence of an increased use, for example, in the arm muscles. The germ material is derived from the parents, and, if it is simply stored in the individual, how could an acquired variation affect it? If an individual lose a limb his offspring will not be without a corresponding limb, for the hereditary material is in the reproductive organs, and it is impossible to believe that the loss of the limb can remove from the hereditary material in the reproductive glands just that part of the germ plasm which was designed for the production of the limb. So, too, if the germ plasm is simply stored in the individual, it is impossible to conceive any way that it can be affected by the conditions around the individual in such a way as to explain the inheritance of acquired variations. If acquired variations do not affect the germ plasm they cannot be inherited, and if the germ plasm is only a bit of protoplasmic substance handed down from generation to generation, we can not believe that acquired variations can influence it.

From such considerations as these have arisen two quite different views among biologists; and, while it is not our purpose to deal with disputed points, these views are so essential to our subject that they must be briefly referred to. One class of biologists adhere closely to the view already outlined, and insist for this reason that acquired variations can not under any conditions be inherited. They insist that all inherited variations are congenital, and due therefore to direct variations in the germ plasm, and that all instances of seeming inheritance of acquired variations are capable of other explanation. The other school is equally insistent that there are abundant instances of the inheritance of acquired characters, claiming that these proofs are so strong as to demand their acceptance. Hence this class of biologists insist that the explanation of heredity given as a simple handing down from generation to generation of a germ plasm is not complete, and that while it is doubtless the foundation of heredity, it must be modified in some way so as to admit of the inheritance of acquired characters. There is no question that has excited such a wide interest in the biological world during the last fifteen years as this one of the inheritance of acquired characters. Until about 1884 the question was not seriously raised. Heredity was known to be a fact, and it was believed that while congenital characters are more commonly inherited, acquired characters may also frequently be handed down from generation to generation. The facts which we have noted of the continuity of germ plasm have during the last fifteen years led many biologists to deny the possibility of the latter. The debate which arose has continued vigorously, and can not be regarded as settled at the present time. One result of this debate is clear. It has been shown beyond question that while the inheritance of congenital characters is the rule, the inheritance of acquired characters is at all events unusual. At the present time many naturalists would be inclined to think that the balance of evidence indicates that under certain conditions certain kinds of acquired characters may be inherited, although this is still disputed by others. Into this discussion we cannot enter here. The reason for referring to it at all is, however, evident. We are searching for nature's method of building machines. It is perfectly clear that variations among animals and plants are the foundations of the successive steps in advance made in this machine building, but of course only such variations as can be transmitted to posterity can serve any purpose in this development. If therefore it should prove that acquired characters can not be inherited, then we should no longer be able to look upon the direct influence of the surroundings as a factor in the machine building. We should then have nothing left except the congenital variations produced by sexual union, or the direct variation of the germ plasm as a factor for advance. If, however, it shall prove that acquired characters may even occasionally be inherited, then the direct effect of the environment upon the individual will serve as a decided assistance in our problem.

Here, then, we have before us the factors which have been concerned in the building of the living machine under nature's hands. Reproduction keeps in existence a constantly active, unstable, readily modified organism as a basis upon which to build. Variation offers constantly new modifications of the type, while heredity insures that the modifications produced in the machine by the influences which give rise to the variations shall be permanently fixed.

Method of Machine Building.—Natural Selection. The method by which these factors have worked together to build up the living machines is easily understood in its general aspects, although there are many details as yet unsolved. The general facts connected with the evolution of animals are matters of common knowledge. We need do no more than outline the subject, since it is well understood by all. The basis of the method is natural selection, which acts in this machine building something as follows:

The law of reproduction, as we have seen, produces new individuals with extraordinary rapidity, and as a result more individuals are born than can possibly find sustenance in the world. Hence only a few of the offspring of any animal or plant can live long enough to produce offspring in turn. The many must die that the few may live; and there is, therefore, a constant struggle among the individuals that are born for food or for room in the world. In this struggle for existence of course the weakest will go to the wall, while those that are best adapted for their place in life will be the ones to get food, live, and reproduce their kind. This is at all events true among the lower animals, although with mankind the law hardly applies. Now, among the individuals that are born there will be no two exactly alike, since variations are universal, many of which are congenital and thus born with the individual and transmitted by inheritance. Clearly enough those animals that have a variation which makes them a little better adapted for the struggle will be the ones to live and hence to produce offspring, while those without such advantage will be the ones to die. We may suppose, for example, that some of the individuals had longer necks than the average. In time of scarcity of food these individuals would be able to get food that the short-necked individuals could not reach. Hence in times of famine the long-necked individuals would be the ones to survive. Now if this peculiarity were a congenital variation it would be already represented in the germ plasm, and consequently it would be inherited by the next generation. The short-necked individuals being largely destroyed in this struggle for food, it would follow that the next generation would be a little better off than the last, since all would inherit this tendency toward a long neck. A few generations would then see the disappearance of all individuals which did not show either this or some other corresponding advantage, and in this way the lengthened neck would be added permanently as a part of the machine. When this time came this peculiarity would no longer give its possessors any advantage over its rivals, since all would possess it. Now, therefore, some new variation would in the same way determine which animals should live and which should die in the struggle, and in time a new modification would be added to the machine. And thus this process continues, one variation after another being added, until the machine is slowly built into a more and more complicated structure, always active but with a constantly increasing efficiency. The construction is a natural one. A mixing of germ plasm in sexual reproduction or some other agencies produce congenital variations; natural selection acting upon the numerous progeny selects the best of the new variations, and heredity preserves and hands them down to posterity.

All students of whatever school recognize the force of this principle and look upon natural selection as an efficient agency in machine building. It is probably the most fundamental of the external laws that have guided the process. There are, however, certain other laws which have played a more or less subordinate part. The chief of these are the influence of migration and isolation, and the direct influence of the environment. Each of these laws has its own school of advocates, and each has been given by its advocates the chief role in the process of machine building.

Migration and Isolation.—The production of the various types of machines has been undoubtedly facilitated by the migrations of animals and the isolation of different groups of descendants from each other by various natural barriers. The variations which occur in organisms are so great that they would sometimes run into abnormal structures were it not for the fact that sexual reproduction constantly tends to reduce them. In an open country where animals and plants interbreed freely, it will commonly happen that individuals with certain peculiarities will mate with others without such peculiarities, and the offspring will therefore inherit the peculiarity not in increased degree but in decreased degree. This constant interbreeding of individuals will tend to prevent the formation of many modifications in the machine which become started by variations. Now plainly if some such individuals, with a peculiar variation, should migrate into a new territory or become isolated from their relatives which do not have similar variations, these individuals will be obliged to breed with each other. The result will be that the next generation, arising thus from two parents each of which shows the same variation, will show it also in equal or increased degree. Migrations and isolations will thus tend to fix in the machine variations which sexual union or other influences inaugurate. Now in the history of the earth's surface there have been many changes which tend to bring about such migration and isolations, and this factor has doubtless played a more or less important part in the building of the machines. How great a part we cannot say, nor is it necessary for our purpose to decide; for in all these cases the machine building has only been the result of the hereditary transmission of congenital variation under certain peculiar conditions. The fundamental process is the same as already considered, only the details of its working being in question.

Direct Influence of the Environment.—Under this head we have a subject of great importance. It is an undoubted fact that the environment has a very decided effect upon the machine. These direct effects of the environment are very positive and in great variety. The tropical sun darkens the human skin; cold climate stunts the growth of plants; lack of food dwarfs all animals and plants, and hundreds of other similar examples could be selected. Another class of similar influences are those produced by use and disuse. Beyond question the use of an organ tends to increase its size, and disuse to decrease it. Combats of animals with each other tend to increase their strength, flight from enemies their running powers, etc.

Now all these effects are direct modifications of the machine, and if they are only transmitted to following generations so as to become permanent modifications, they will be most important agencies in the machine building. If, on the other hand, they are not transmitted by heredity, they can have no permanent effect. We have here thus again the problem of the inheritance of acquired characters. We have already noticed the uncertainty surrounding this subject, but the almost universal belief in the inheritance of such characters requires us to refer to it again. It is uncertain whether such direct effects have any influence upon the offspring, and therefore whether they have anything to do with this machine building. Still, there are many facts which point strongly in this direction. For example, as we study the history of the horse family we find that an originally five-toed animal began to walk more and more on its middle toe, in such a way that this toe received more and more use, while the outer toes were used less and less. Now that such a habit would produce an effect upon the toes in any generation is evident; but apparently this influence extended from generation to generation, for, as the history of the animals is followed, it is found that the outer toes became smaller and smaller with the lapse of ages, while the middle one became correspondingly larger, until there was finally produced the horse with its one toe only on each foot. Now here is a line of descent or machine building in the direct line of the effects of use and disuse, and it seems very natural to suppose that the modification has been produced by the direct effect of the use of the organs. There are many other similar instances where the line of machine building has been quite parallel to the effects of use and disuse. If, therefore, acquired characters can be inherited to any extent, we have, in the direct influences of the environment an important agency in machine building. This direct effect of the conditions is apparently so manifest that one school of biologists finds in it the chief cause of the variations which occur, telling us that the conditions surrounding the organism produce changes in it, and that these variations, being handed down to subsequent generations, constitute the basis of the development of the machine. If this factor is entirely excluded, we are driven back upon the natural selection of congenital variations as the only kind of variations which can permanently effect the modification of the machine.

Consciousness.—It may be well here to refer to one other factor in the problem, because it has somewhat recently been brought into prominence. This factor is consciousness on the part of the animal. Among plants and the lower animals this factor can have no significance, but consciousness certainly occurs among the higher animals. Just when or how it appeared are questions which are not answered, and perhaps never will be. But consciousness, after it had once made its appearance, became a controlling factor in the development of the machine. It must not be understood by this that animals have had any consciousness of the development of their body, or that they have made any conscious endeavours to modify its development. This has not always been understood. It has been frequently supposed that the claim that consciousness has an influence upon the development of an animal means that the animal has made conscious efforts to develop in certain directions. For example, it has been suggested that the tiger, conscious of the advantage of being striped, had a desire to possess stripes, and the desire caused their appearance. This is absurd. Consciousness has been a factor in the development of the machine, but an indirect one. Consciousness leads to effort, and effort has a direct influence in development. For example, an animal is conscious of hunger, and this leads to efforts on his part to obtain food. His efforts to obtain food may lead to migration or to the adoption of new kinds of food or to conflicts with various kinds of rivals, and all of these efforts are potent factors in determining the direction of development. Consciousness, again, may lead certain animals to take pleasure in each other's society, or to recognize that in mutual association they have protection against common enemies. Such a consciousness will give rise to social habits, and social habits are a very potent factor in determining the direction in which the inherited variations will tend; not, perhaps, because it effects the variations themselves, but rather because it determines which variations among the many shall be preserved and which rejected by natural selection. Consciousness may lead the antelope to recognize that he has no chance in a combat with a lion, and this will induce him to flee. The habit of flight would then develop the power of flight, not because the antelope desired such power, but because the animals with variations which gave increased power of flight would be the ones to escape the lion, while the slower ones would die without offspring. Thus consciousness would indirectly, though not directly, result in the lengthening of the legs of the animal and in the strengthening of his running muscles. Beyond a doubt this factor of consciousness has been a factor of no little moment in the development of the higher types of organic machines. We can as yet only dimly understand its action, but it must hereafter be counted as one of the influences in the evolution of the living machine.

But, after all, these are only questions of the method of the action of certain well demonstrated, fundamental factors. Whether by natural selection, or by the inheritance of acquired characters produced by the environment, or whether by the effect of isolation of groups of individuals, the machine building has always been produced in the same way. A machine, either through the direct influence of the environment, or as a result of sexual combination of germ plasm, shows a variation from its parents. This variation proves of value to its possessor, who lives and transmits it permanently to posterity. Thus step by step, one part is added to another, until the machine has grown into the intricately adapted structure which we call the animal or plant. This has been nature's method of building machines, all based upon the three properties possessed by the living cell—reproduction, variation, and heredity.

Summary of Nature's Power of Building Machines.—Let us now notice the position we have reached. Our problem in the present chapter has been to find out whether nature possesses forces adequate to explain the building of machines with their parts accurately adapted to each other so as to act harmoniously for certain ends. Astronomy has shown that she has forces for the building of worlds; geology, that she has forces for making mountain and valley; and chemistry, that she has forces for building chemical compounds. But the organism is neither a world, nor a mass of matter, nor a chemical compound. It is a machine. Has nature any forces for machine building? We have found that by the use of the three factors, reproduction, variation, and heredity, nature is able to produce a machine of ever greater and greater complexity, with the parts all adapted to each other. Now the difference between a machine and a mass of matter is simply in the adaptation of parts to act harmoniously for definite ends. Hence if we are allowed these three factors, we can say that nature does possess forces adequate to the manufacture of machines. These forces are not chemical forces, and the construction of the machine has thus been brought about by forces entirely different from those which produced the chemical molecule.

But we have plainly not reached the bottom of the matter in our attempt to explain the machinery of living things. We have based the whole process upon three factors. Reproduction, variation, and heredity are the properties of all living matter; but they are not, like gravity and chemism, universal forces of nature. They occur in living organisms only. Why should they occur in living organisms, and here alone? These three properties are perhaps the most marvellous properties of nature; and surely we have not finished our task if we have based the whole process of machine building upon these mysterious phenomena, leaving them unintelligible. We must therefore now ask whether we can proceed any farther and find any explanation of these fundamental powers of the living machine.

It must be confessed that here we are at present forced to stop. We can proceed no further with any certainty, or even probability. We may say that variation and heredity are only phases of reproduction, and reproduction is a property of the living cell. We may say that this power of reproduction is dependent upon the power of assimilation and growth, for cell division is a result of cell growth. We may further say that growth and assimilation are chemical processes resulting from the oxidation of food, and that thus all of these processes are to be reduced to chemical forces. In this way we may seem to have a chemical foundation for life phenomena. But clearly this is far from satisfactory. In the first place, it utterly fails to explain why the living cell has these properties, while no other body possesses them, nor why they are possessed by living protoplasms alone, ceasing instantly with death. Indeed it does not tell us what death can be. Secondly, it utterly fails to explain the marvels of cell division with resulting hereditary transmission. For all this we must fall back upon the structure of protoplasm, and say that the cell machinery is so adjusted that the machine, when acting as a whole, is capable of transforming the energy of chemical composition in certain directions. These fundamental properties are then the properties of the cell machine just as surely as printing is the property of the printing press. We can no more account for the life phenomena by chemical powers than we can for printing by chemical forces manifested in the burning of the coal in the engine room. To be sure, it is the chemical forces in the engine room that furnishes the energy, but it is the machinery of the press that explains the printing. So, while chemical forces supply life energy, it is the cell machinery that must explain the fundamental living factors. So long as this machine is intact it can continue to run and perform its duties. But it is a very delicate machine and is easily broken. When it is broken its activities cease. A broken machine can not run. It is dead. In short, we come back once more to the idea of the machinery of protoplasm, and must base our understanding of its properties upon its structure.

It is proper to state that there are still some biologists who insist that the ultimate explanation of protoplasm is purely chemical and that life phenomena may be manifested in mixtures of compounds that are purely physical mixtures and not machines. It is claimed that much of this cell structure described above is due to imperfection in microscopic methods and does not really exist in living protoplasm, while the marvellous activities described are found only in the highly organized cell, but do not belong to simple protoplasm. It is claimed that simple protoplasm consists of a physical mixture of two different compounds which form a foam when thus mixed, and that much of the described structure of protoplasm is only the appearance of this foam. This conception is certainly not the prevalent one to-day; and even if it should be the proper one, it would still leave the cell as an extremely complicated machine. Under any view the cell is a mechanism and must be resolved into subordinate parts. It may be uncertain whether these subordinate parts are to be regarded simply as chemical compounds physically mixed, or as smaller units each of which is a smaller mechanism. At all events, at the present time we know of no such simple protoplasm capable of living activities apart from machinery, and the problem of explaining life, even in the simplest form known, remains the problem of explaining a mechanism.

The Origin of the Cell Machine.—We have thus set before us another problem, which is after all the fundamental one, namely, to ask whether we can tell anything of nature's method of building the protoplasmic machine. The building of the higher animal and plant, as we have seen, is the result of the powers of protoplasm; but protoplasm itself is a machine. What has been its history?

We must first notice that no notion of chemical evolution helps us out. It has been a favourite thought with some that the origin of the first living thing was the result of chemical evolution. As the result of physical forces there was produced, from the original nebulous mass, a more and more complicated system until the world was formed. Then chemical phenomena became more and more complicated until, with the production of more and more complicated compounds, protoplasm was finally produced. A few years ago, under the impulse of the idea that protoplasm was a compound, or at least a simple mixture of compounds, this thought of protoplasm as the result of chemical evolution was quite significant. Physical forces, chemical forces, and vital forces, explain successively the origin of worlds, protoplasm, and organisms. This conception has, however, no longer much significance. We know of no such living chemical compound apart from cell machinery. A new conception of protoplasm has arisen which demands a different explanation of its origin. Since it is a machine rather than a compound, mechanical rather than chemical forces are required for its explanation.

Have we then any suggestion as to the method of the origin of this protoplasmic machine? Our answer must, at the present, be certainly in the negative. The complexity of the cell tells us plainly that it can not be the ultimate living substance which may have arisen from chemical evolution. It is made up of parts delicately adapted to act in harmony with each other, and its activity depends upon the relation of these parts. Whatever chemical forces may have accomplished, they never could have combined different bodies into linin, centrosomes, chromosomes, etc., which, as we have seen, are the basis of cell life. To account for this machine, therefore, we are driven to assume either that it was produced by some unknown intelligent power in its present condition of complex adjustment, or to assume that it has had a long history of building by successive steps, just as we have seen to be the case with the higher organisms. The latter assumption is, of course, in harmony with the general trend of thought. To-day protoplasm is produced only from other protoplasm; but, plainly, the first protoplasm on the earth must have had a different origin. We must therefore next look for facts which will enable us to understand its origin. We have seen that the animal and plant machines have been built up from the simple cell as the result of its powers acting under the ordinary conditions of nature. Now, in accordance with this general line of thought, we shall be compelled to assume that previous to the period of building machinery which we have been considering, there was another period of machine building during which this cell machine was built by certain natural forces.

But here we are forced to stop, for nothing which we yet know gives even a hint as to the method by which this machine was produced. We have, however, seen that there are forces in nature efficient in building machines, as well as those for producing chemical compounds; and this, doubtless, suggests to us that there may be similar forces at work in building protoplasm. If we can find natural forces by which the simplest bit of living matter can be built up into a complicated machine like the ox, with its many delicately adjusted parts, it is certainly natural to imagine that the same forces may have built this simpler machine with which we started. But such a conclusion is for a simple reason impossible. We have seen that the essential factor in this machine building is reproduction, with the correlated powers of variation and heredity. Without these forces we could not have advanced in this machine building at all. But these properties are themselves the result of the machinery of protoplasm. We have no reason for thinking that this property of reproduction can occur in any other object in nature except this protoplasmic machine. Of course, then, if reproduction is the result of the structure of protoplasm we can not use this factor in explaining the origin of this protoplasm. The powers of the completed machine can not be brought forward to account for its origin. Thus the one fundamental factor for machine building is lacking, and if we are to explain nature's method of producing protoplasm from simpler structures, we must either suppose that the parts of the cell are capable of reproduction and subject to heredity, or we must look for some other method. Such a road has however not yet been found, nor have we any idea in what direction to look. But the fact that nature has methods of machine building, as we have seen, may hold out the possibility that some day we may discover her method of building this primitive living machine, the cell.

It is useless to try to go further at present. The origin of living matter is shrouded in as great obscurity as ever. We must admit that the disclosures of the modern microscope have complicated rather than simplified this problem. While a few years ago chemists and biologists were eagerly expecting to discover a method of manufacturing a bit of living matter by artificial means, that hope has now been practically abandoned. The task is apparently hopeless. We can manipulate chemical forces and produce an endless series of chemical compounds. But we can not manipulate the minute bits of matter which make up the living machine. Since living matter is made of the adjustment of these microscopic parts of matter, we can not hope to make a bit of living matter until we find some way of making these little parts and adjusting them together. Most students of protoplasm have therefore abandoned all expectation of making even the simplest living thing. We are apparently as far from the real goal of a natural explanation of life as we were before the discovery of protoplasm.

General Summary.—It is now desirable to close this discussion of seemingly somewhat unconnected topics by bringing them together in a brief summary. This will enable us to see more clearly the position in which science stands to-day upon this matter of the natural explanation of living phenomena, and to picture to ourselves more concisely our knowledge of the living machine.

The problem we have set before us is to find out to what extent it is possible to account for vital phenomena by the application of ordinary natural laws and forces, and therefore to find out whether it is necessary to assume that there are forces needed to explain life which are different from those found in other realms of nature, or whether vital forces are all correlated with physical forces. It has been evident at a glance that the living body is a machine. Like other machines it consists of parts adjusted to each other for the accomplishment of definite ends, and its action depends upon the adjustment of its parts. Like other machines, it neither creates nor destroys energy, but simply converts the potential energy of its foods into some form of active energy, and, like other machines, its power ceases when the machine is broken.

With this understanding the problem clearly resolved itself into two separate ones. The first was to determine to what extent known physical and chemical laws and forces are adequate to an explanation of the various phenomena of life. The second was to determine whether there are any known forces which can furnish a natural explanation of the origin of the living machine. Manifestly, if the first of these problems is insolvable, the second is insolvable also.

In the study of the first problem we have reached the general conclusion that the secondary phenomena of life are readily explained by the application of physical and chemical forces acting in the living machine. These secondary phenomena include such processes as the digestion and absorption of food, circulation, respiration, excretion, bodily motion, etc. Nervous phenomena also doubtless come under this head, at least so far as concerns nervous force. We have been obliged, however, to exclude from this correlation the mental phenomena. Mental phenomena can not as yet be measured, and have not yet been shown to be correlated with physical energy. In other words, it has not yet been proved that mental force is energy at all; and if it is not energy, then of course it can not be included in the laws which govern the physical energy of the universe. Although a close relation exists between physical changes in the brain cells and mental phenomena, no further connection has yet been drawn between mental power and physical force. All other secondary phenomena, however, are intelligently explained by the action of natural forces in the machinery of the living organism.

While we have thus found that the secondary phenomena of life are intelligible as the result of the structure of the machine, certain other fundamental phenomena have been constantly forcing themselves upon our attention as a foundation of these secondary activities. The power of contraction, the power of causing certain kinds of chemical change to occur which result in metabolism, the property of sensibility, the property of reproduction—these are fundamental to all living activity, and are, after all, the real phenomena which we wish to explain. But these are not peculiar to the complicated machines. We can discard all the apparent machinery of the animal or plant and find these properties still developed in the simplest bit of living matter. To learn their significance, therefore, we have turned to the study of the simplest form of matter in which these fundamental properties are manifested. This led us at once to the study of the so-called protoplasm, for protoplasm is the simplest known form of matter that is alive. Protoplasm itself at first seemed to be a homogeneous body, and was looked upon as a chemical compound of high complexity. If this were true its properties would depend upon its composition and would be explained by the action of chemical forces. Such a conception would have quickly solved the problem, for it would reduce living properties to chemical powers. But the conception proved to be delusive. Protoplasm, at least the simplest form known to possess the fundamental life properties, soon showed itself to be no chemical compound, but a machine of wonderful intricacy.

The fundamental phenomena of life and of protoplasm have proved to be both chemical and mechanical. Metabolism is the result of the oxidation of food, and motion is an instance of transference of force. Our problem then resolved itself into finding the power that guides the action of these natural forces. Food will not undergo such an oxidation except in the presence of protoplasm, nor will the phenomena of metabolism occur except in the presence of living protoplasm. Clearly, then, the living protoplasm contains within itself the power of guiding this play of chemical force in such a way as to give rise to vital phenomena, and our search must be not for chemical force but for this guiding principle. Our study of protoplasm has told us clearly enough that we must find this guiding principle in the interaction of the machinery within the protoplasm. The microscope has told us plainly that these fundamental principles are based upon machinery. The cell division (reproduction) is apparently controlled by the centrosomes; the heredity by the chromosomes; the constructive metabolism by the nucleus in general, while the destructive metabolism is also seated in the cell substance outside the nucleus. Whether these statements are strictly accurate in detail does not particularly affect the general conclusion. It is clearly enough demonstrated that the activities of the protoplasmic body are dependent upon the relation of its different parts. Although we have got rid of the complicated machinery of the organism in general, we are still confronted with the machinery of the cell.

But our analysis can not, at present, go further. Our knowledge of this machine has not as yet enabled us to gain any insight as to its method of action. We can not yet conceive how this machine controls the chemical and physical forces at its disposal in such a way as to produce the orderly result of life. The strict correlation between the forces of the physical universe and those manifested by this protoplasm tells us that a transformation of energy occurs within it, but of the method of that transformation we as yet know nothing. Irritability, movement, metabolism, and reproduction appear to be not chemical properties of a compound, but mechanical properties of a machine. Our mechanical analysis of the living machine stops short before it reaches any foundation in the chemical forces of nature.

It is thus clearly apparent that the phenomena of life are dependent upon the machinery of living things, and we have therefore the second question of the origin of this machinery to answer. Chemical forces and mechanical forces have been laboriously investigated, but neither appear adequate to the manufacture of machines. They produce only chemical compounds and worlds with their mountains and seas. The construction of artificial machines has demanded intelligence. But here is a natural machine—the organism. It is the only machine produced by natural methods, so far as we know; and we have therefore next asked whether there are, in nature, simple forces competent to build machines such as living animals and plants?

In pursuance of this question we have found that the complicated machines have been built out of the simpler ones by the action of known forces and laws. The factors in this machine building are simply those of the fundamental vital properties of the simplest protoplasmic machine. Reproduction, heredity, and variation, acting under the ever-changing conditions of the earth's surface, are apparently all that are needed to explain the building of the complex machines out of the simpler ones. Nature has forces adequate to the building of machines as well as forces adequate to the formation of chemical compounds and worlds.

But here again we are unable to base our explanation upon chemical and physical forces. Reproduction, heredity, and variation are properties of the cell machine, and we are therefore thrown back upon the necessity of explaining the origin of this machine. Can we find a mechanical or chemical explanation of the origin of protoplasm? A chemical explanation of the cell is impossible, since it is not a chemical compound, but a piece of mechanism. The explanation given for the origin of animals and plants is also here apparently impossible. The factors upon which that explanation depended are factors of this completed machine itself, and can not be used to explain its origin. We are left at present therefore without any foundation for further advance. The cells must have had a history of construction, but we do not as yet conceive any forces which may be looked upon as contributing to that history. Whether life phenomena can be manifested by any mixture of compounds simpler than the cell we do not yet know.

The great problems still remaining for solution, which have hardly been touched by modern biology in all its endeavours to find a mechanical explanation of the living machine, are, therefore, three. First, the relation of mentality to the general phenomena of the correlation of force; second, the intelligible understanding of the mechanism of protoplasm which enables it to guide the blind chemical and physical forces of nature so as to produce definite results; third, the kind of forces which may have contributed to the origin of that simplest living machine upon whose activities all vital phenomena rest—the living cell.